T0313762

DESIGN AND REALIZATIONS OF MINIATURIZED FRACTAL RF AND MICROWAVE FILTERS

DESIGN AND REALIZATIONS OF MINIATURIZED FRACTAL RF AND MICROWAVE FILTERS

PIERRE JARRY
Professor, Bordeaux University

JACQUES BENEAT
Associate Professor, Norwich University

A JOHN WILEY & SONS, INC., PUBLICATION

Published by John Wiley & Sons, Inc., Hoboken, New Jersey.
Published simultaneously in Canada.

For general information on our other products and services or for technical support, please contact our Customer Care Department within the United States at (800) 762-2974, outside the United States at (317) 572-3993 or fax (317) 572-4002.

Wiley also publishes its books in a variety of electronic formats. Some Content that appears in print may not be available in electronic formats. For more information about Wiley products, visit our web site at www.wiley.com.

***Library of Congress Cataloging-in-Publication Data*:**

Jarry, Pierre, 1946–
Design and realizations of miniaturized fractal RF and microwave filters / Pierre Jarry, Jacques Beneat.
p. cm.
Includes bibliographical references and index.
ISBN 978-0-470-48781-5 (cloth)
1. Microwave filters. 2. Radio filters. 3. Miniature electronic equipment. I. Beneat, Jacques, 1964- II. Title.
TK7872.F5J372 2009
621.381′3224–dc22

2009014024

10 9 8 7 6 5 4 3 2 1

CONTENTS

FOREWORD ix

PREFACE xi

1 MICROWAVE FILTER STRUCTURES 1

1.1 Background / 2
1.2 Cavity Filters / 3
1.2.1 Evanescent-Mode Waveguide Filters / 3
1.2.2 Coupled Cavity Filters / 5
1.2.3 Dielectric Resonator Filters / 9
1.2.4 E-Plane Filters / 12
1.3 Planar Filters / 13
1.3.1 Semi-lumped Filters / 14
1.3.2 Planar Transmission Line Filters / 14
1.4 Planar Filter Technology / 19
1.4.1 Microstrip Technology / 19
1.4.2 Coplanar Technology / 20
1.4.3 Suspended Substrate Stripline Technology / 20
1.4.4 Multilayer Technology / 21
1.5 Active Filters / 22
1.6 Superconductivity or HTS Filters / 23
1.7 Periodic Structure Filters / 24
1.8 SAW Filters / 25

1.9 Micromachined Filters / 27
1.10 Summary / 28
References / 29

2 IN-LINE SYNTHESIS OF PSEUDO-ELLIPTIC FILTERS 35

2.1 Introduction / 35
2.2 Approximation and Synthesis / 36
2.3 Chebyshev Filters / 36
2.4 Pseudo-elliptic Filters / 37
2.4.1 Pseudo-elliptic Characteristic Function / 38
2.4.2 Pseudo-elliptic Transfer Functions / 40
2.4.3 Cross-Coupled and In-Line Prototypes for Asymmetrical Responses / 41
2.4.4 Analysis of the In-Line Prototype Elements / 45
2.4.5 Synthesis Algorithm for Pseudo-elliptic Lowpass In-line Filters / 49
2.4.6 Frequency Transformation / 50
2.5 Prototype Synthesis Examples / 52
2.6 Theoretical Coupling Coefficients and External Quality Factors / 59
References / 61

3 SUSPENDED SUBSTRATE STRUCTURE 63

3.1 Introduction / 63
3.2 Suspended Substrate Technology / 64
3.3 Unloaded Quality Factor of a Suspended Substrate Resonator / 66
3.4 Coupling Coefficients of Suspended Substrate Resonators / 69
3.4.1 Input/Output Coupling / 69
3.4.2 Coupling Between Resonators / 69
3.5 Enclosure Design Considerations / 73
References / 75

4 MINIATURIZATION OF PLANAR RESONATORS USING FRACTAL ITERATIONS 79

4.1 Introduction / 79
4.2 Miniaturization of Planar Resonators / 81
4.3 Fractal Iteration Applied to Planar Resonators / 82

4.4 Minkowski Resonators / 84
4.4.1 Minkowski's Fractal Iteration / 84
4.4.2 Minkowski Square First-Iteration Resonator / 85
4.4.3 Minkowski Rectangular First-Iteration Resonator / 86
4.4.4 Minkowski Second-Iteration Resonators / 88
4.4.5 Sensitivity of Minkowski Resonators / 90
4.5 Hibert Resonators / 92
4.5.1 Hilbert's Curve / 92
4.5.2 Hilbert Half-Wavelength Resonator / 93
4.5.3 Unloaded Quality Factor of Hilbert Resonators / 94
4.5.4 Sensitivity of Hilbert Resonators / 96
References / 98

5 DESIGN AND REALIZATIONS OF MEANDERED LINE FILTERS 101

5.1 Introduction / 102
5.2 Third-order Pseudo-elliptic Filters with Transmission Zero on the Right / 102
5.2.1 Topology of the Filter / 102
5.2.2 Synthesis of the Bandpass Prototype / 103
5.2.3 Defining the Dimensions of the Enclosure / 103
5.2.4 Defining the Gap Distances / 105
5.2.5 Further Study of the Filter Response / 107
5.2.6 Realization and Measured Performance / 110
5.2.7 Sensitivity Analysis / 114
5.3 Third-order Pseudo-elliptic Filters with Transmission Zero on the Left / 117
5.3.1 Topology of the Filter / 117
5.3.2 Synthesis of the Bandpass Prototype / 118
5.3.3 Defining the Dimensions of the Enclosure / 118
5.3.4 Defining the Gap Distances / 118
5.3.5 Realization and Measured Performance / 122
5.3.6 Sensitivity Analysis / 125
References / 127

6 DESIGN AND REALIZATIONS OF HILBERT FILTERS 129

6.1 Introduction / 129
6.2 Design of Hilbert Filters / 130
6.2.1 Second-Order Chebyshev Responses / 130

6.2.2 Third-Order Chebyshev Responses / 134
6.2.3 Third-Order Pseudo-elliptic Response / 136
6.2.4 Third-Order Pseudo-elliptic Responses Using Second-Iteration Resonators / 144
6.3 Realizations and Measured Performance / 148
6.3.1 Third-Order Pseudo-elliptic Filter with Transmission Zero on the Right / 149
6.3.2 Third-Order Pseudo-elliptic Filter with Transmission Zero on the Left / 154
References / 159

7 DESIGN AND REALIZATION OF DUAL-MODE MINKOWSKI FILTERS 161

7.1 Introduction / 161
7.2 Study of Minkowski Dual-Mode Resonators / 162
7.3 Design of Fourth-Order Pseudo-elliptic Filters with Two Transmission Zeros / 166
7.3.1 Topology / 166
7.3.2 Synthesis / 167
7.3.3 Design Steps / 168
7.4 Realization and Measured Performance / 176
References / 182

APPENDIX 1: Equivalence Between *J* and *K* Lowpass Prototypes 183

APPENDIX 2: Extraction of the Unloaded Quality Factor of Suspended Substrate Resonators 189

INDEX 193

FOREWORD

Microwave and RF filtering technologies are key to controlling the spectrum of signals and tackling interference issues in many wireless systems, including mobile communications. For most current wireless systems, the physical volume of the RF front end is dominated by the size of the necessary filters. Since the trend for future wireless systems is toward smaller and lighter, there has been an increasing demand for the miniaturization of microwave and RF filters. To this end, I was very delighted to read the manuscript of the *Design and Realizations of Miniaturized Fractal RF and Microwave Filters*, the first book specified on this subject. The book is based on the authors' research work, and both of them, Professors Pierre Jarry and Jacques Beneat, are well-known experts in this subject field. With an interesting introduction to the history of fractals, the book shows how these fascinating fractal constructions, including Hilbert and Minkowski fractal iterations, can be applied to microwave filter miniaturizing. Numerous design examples are presented, some of which are dealt with a great detail. The book has been written in such a way that it helps readers to gain the necessary knowledge and skills to design and realize miniaturized fractal microwave and RF filters. Innovative applications of fractals for the miniaturization of microwave filters require imagination, and this book demonstrates the art of this engineering approach. The book will be extremely useful to practitioners engaged in microwave filter design, as well as to students and researchers on the subject.

JIA-SHENG HONG

Heriot-Watt University at Edinburgh
The United Kingdom

PREFACE

Microwave and RF filters play an important role in communication systems, and due to the proliferation of radar, satellite, and mobile wireless systems, there is a need for design methods that can satisfy the ever-increasing demand for accuracy, reliability, and fast development time. This book is dedicated to the synthesis and fabrication of suspended substrate fractal filters with symmetrical and asymmetrical frequency characteristics in the C, X, and Ku bands. The method is general enough to be applied to other technologies and frequency bands. The book is intended for researchers and RF and microwave engineers but is also suitable for an advanced graduate course in the subject area of miniaturized RF and microwave filters. Furthermore, it is possible to choose material from the book to supplement traditional courses in microwave filter design.

The shape of planar resonators is based on fractal iterations. It is seen that the use of fractal iterations to shape the resonator is repeatable and follows a systematic method. It is shown that for a given resonant frequency, a resonator based on fractal iterations can become significantly smaller than conventional planar resonators. It is also shown that by increasing the number of fractal iterations, the resonators can be reduced further. The studies presented in this book show that the quality factor of the resonator can be increased by increasing the fractal iteration number. Compared to other advanced planar resonators, besides the obvious reduction in size, fractal filters present very good spurious responses. They occur well beyond twice the center frequency of the filter. It is also shown that by increasing the number of fractal iterations, the resonators can produce transmission zeros. Therefore, pseudo-elliptic responses can be implemented. Pseudo-elliptic responses allow further reduction in the size of the filters, since for the same required selectivity, the order of the filter can be reduced by the addition of transmission zeros close to the passband. If improved selectivity is needed on only one side of the filter, a transmission zero can be used without its symmetrical counterpart. This results in asymmetrical pseudo-elliptic responses that are often more difficult to implement. In this book, most realizations are provided for asymmetrical pseudo-elliptic responses.

Another technique for reducing the size of a filter is to use coupled modes. Studies in this book show how resonators based on Minkowski fractal iterations are modified by introducing dissymmetry in their shape so that two resonant modes can be decoupled. In this manner, fourth-order pseudo-elliptic responses can be obtained with only two dual-mode resonators, half the number needed using single-mode resonators. This results in very compact filters with improved selectivity and spurious responses.

The design technique is based on the general coupled resonators network and coupling matrix technique. After synthesizing a lowpass pseudo-elliptic prototype circuit based on a new admittance extraction technique, the theoretical coupling coefficients and external quality factors can be computed. Then the theoretical coupling coefficients and external quality factors are matched to the experimental coupling coefficients and external quality factors of the structure using a commercial electromagnetic simulator. All the enclosures were fabricated in-house (at the university), while the substrate and planar resonators were fabricated through Atlantec, a general printed circuit board company not specializing in microwave devices. The acceptable filter responses obtained for such low-cost fabrication techniques show that fabrication of fractal shapes is not a major issue. One should also note that the filters were designed to operate at frequencies often above 10 GHz and require no tuning.

The book is divided into seven chapters:

Chapter 1 provides a rather exhaustive overview of current microwave structures used for filtering applications. It provides descriptions of the various technologies, filtering characteristics obtainable, and suitable applications. This chapter is a good reference on the state of the art of RF and microwave filters.

In Chapter 2 we present an original extraction procedure called in-line synthesis to provide the elements of an in-line lowpass network that can produce symmetrical or asymmetrical pseudo-elliptic filtering characteristics. This technique is easier than the challenging chain matrix extraction procedure used for cross-coupled networks. The bandpass prototype network is defined using a frequency transformation technique suitable for asymmetrical responses. Several step-by-step synthesis examples are provided. Finally, the theoretical coupling matrix and external quality factor corresponding to third-order pseudo-elliptic filters with one transmission zero are recalled.

Chapter 3 provides background material regarding suspended substrate technology. A method for extracting the experimental coupling coefficients using a full-wave electromagnetic simulator is presented. Some sensitivity analyses of the effects of the various physical parameters used in suspended substrate technology are given. This technology has several advantages and is used to describe the design method of fractal filters. However, the method is general enough to be applied to other technologies.

In Chapter 4 we present a brief history of fractals. It is shown that the overall size of planer resonators can be reduced by changing their shape. Introductory examples of resonators based on a straight line, a line shaped in the form of a square, and a more general meandered line show that size can be reduced in order to achieve the same resonant frequency. We then focus on Minkowski and Hilbert fractal iterations to provide a systematic method for designing the resonators. The fractal resonators have several interesting features. For the same resonant frequency, the size of the resonator must be reduced when increasing the fractal iteration number. The fractal resonators can provide better stopband rejections, produce transmission zeros, and at times can provide better quality factors.

In Chapter 5 we introduce the design and realizations of miniaturized resonator filters. The filters use half-wavelength meandered line resonators. The responses are very comparable to those of conventional half-wavelength line resonator filters. However, besides the obvious reduction in size, the meandered line filters have better spurious responses and improved selectivity. We present two realizations. The first has a third-order pseudo-elliptic characteristic with a transmission zero on the right side of the passband, while the second has a transmission zero on the left side. The enclosures were fabricated in-house, while the substrate and planar resonators were fabricated by Atlantec. It is seen that the measured responses are slightly shifted toward higher frequencies but are very good for such low-cost fabrication. Sensitivity analyses are given to show the effects of various physical parameters on the response of the filters.

Chapter 6 provides the design technique of miniaturized filters using Hilbert fractal iteration resonators. The chapter covers the topology to be used so that the resonators can produce electric or magnetic couplings to create transmission zeros on the left or on the right of the passband. Two realizations are presented. The first is a third-order pseudo-elliptic filter with a transmission zero on the right of the passband, and the second has a transmission zero on the left. The performances measured agree well with the simulations, except for a slight shift toward higher frequencies. However, the filters were fabricated using the same low-cost solutions as those used for meandered line filters.

In Chapter 7 we provide design technique for miniaturized filters using dual-mode Minkowski fractal iteration resonators. By introducing a dissymmetry in the geometry of the resonator (e.g., a cut in the corner), it is possible to excite two modes of propagation. It is seen that a single dual-mode Minkowski resonator can produce two resonant frequencies and two transmission zeros. Unfortunately, this dual-mode resonator does not exactly fit the structure of the in-line bandpass prototype network of Chapter 2, and some modifications of the synthesis method are needed. The realization of a fourth-order pseudo-elliptic filter with transmission zeros on both sides of the passband is given. The measured performance agrees well with the simulations except for a slight shift toward higher frequencies. However, the filter was fabricated using the same low-cost solutions as those used for meandered line filters.

In summary, the book shows how fractal iterations applied to microwave resonators lead to RF and microwave filters of reduced size, improved selectivity, and stopband rejections. The dual-mode filter reduces the size of the filter further by using two modes of resonances. In this case it is possible to obtain a fourth-order filter with transmission zeros on both sides of the passband using only two resonators. The results described regarding the synthesis of pseudo-elliptic filters and the use of fractal iterations can be extended to other technologies with better insertion losses, such as micromachining, superconductivity, and LTCC technology. The small size requirements of fractal filters can then offset the high fabrication costs of these technologies.

Acknowledgments

We are most indebted to Jia-Sheng Hong, professor at Heriot-Watt University at Edinburgh in the United Kingdom. His help is gratefully acknowledged. We extend our sincere gratitude to Dr. Inder Bahl, of Cobham–USA, Editor of the *International Journal of RF and Microwave Computer Aided Engineering*. This book could not have been written without his help. We would also like to acknowledge the major contributions of Dr. Engineer Elias Hanna, our research student, whose thesis collaboration has resulted in much of the material in the book. Our sincere gratitude extends to our publisher at Wiley, George Telecki, and to the reviewers for their support in writing the book.

Pierre Jarry wishes to thank his colleagues at the University of Bordeaux and the University of Brest, including Professor Pascal Fouillat and Professor Eric Kerherve, and also Professor Yves Garault at University of Limoges. Finally, he would like to express his deep appreciation to his wife, Roselyne, and to his son, Jean-Pierre, for their tolerance and support.

Jacques Beneat is very grateful to Norwich University, a place conducive to trying and succeeding in new endeavours. He particularly wants to thank Professors Steve Fitzhugh, Ronald Lessard, and Michael Prairie for their support during the writing of the book, and the acting Vice President for Academic Affairs and Dean of the Faculty, Frank Vanecek, and the Dean of the David Crawford School of Engineering, Bruce Bowman, for their encouragements in such a difficult enterprise.

PIERRE JARRY

JACQUES BENEAT

CHAPTER 1

MICROWAVE FILTER STRUCTURES

1.1 Background 2
1.2 Cavity filters 3
1.2.1 Evanescent-mode waveguide filters 3
1.2.2 Coupled cavity filters 5
1.2.3 Dielectric resonator filters 9
1.2.4 E-plane filters 12
1.3 Planar filters 13
1.3.1 Semi-lumped filters 14
1.3.2 Planar transmission line filters 14
1.4 Planar filter technology 19
1.4.1 Microstrip technology 19
1.4.2 Coplanar technology 20
1.4.3 Suspended substrate stripline technology 20
1.4.4 Multilayer technology 21
1.5 Active filters 22
1.6 Superconductivity or HTS filters 23
1.7 Periodic structure filters 24
1.8 SAW filters 25
1.9 Micromachined filters 27
1.10 Summary 28
References 29

Design and Realizations of Miniaturized Fractal RF and Microwave Filters,
By Pierre Jarry and Jacques Beneat

1.1 BACKGROUND

On October 4, 1957, *Sputnik I*, the first man-made satellite, was sent into space by the Soviet Union. For three weeks, the satellite relayed information back to Earth using radio transmissions. Toward the end of the 1960s, *Intelsat I* was the first of a series of satellites to be used for providing new means of communications. By 1971, *Intelsat IV* confirmed the importance of satellites for commercial communications. Since then, the number of satellites sent into space has increased constantly, and today they are an essential part of modern communication networks.

The main element of a communication satellite is the repeater. A *repeater* is used to receive signals on a given carrier frequency (the uplink) from a transmitting station and to forward these signals to a receiving station located somewhere else on Earth using another carrier frequency (the downlink), as shown in Figure 1.1. The simplified diagram of a satellite repeater in Figure 1.2 shows the presence of several filters. These filters are designed for space applications

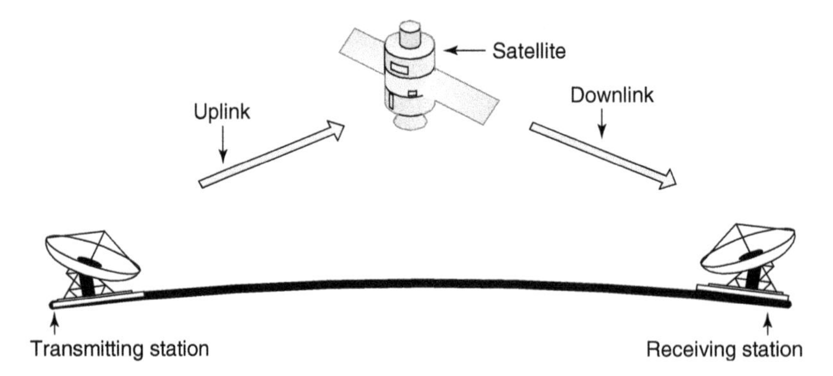

Figure 1.1 Satellite communications between two terrestrial stations.

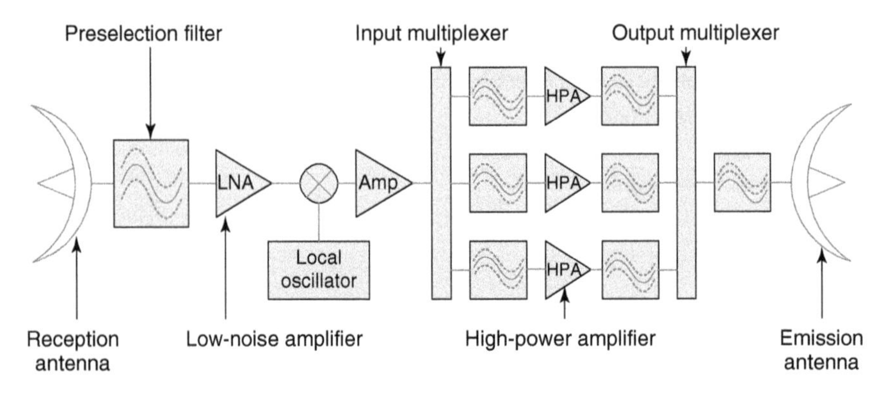

Figure 1.2 Block diagram of a satellite repeater.

and must have very low losses since the energy resources aboard a satellite are limited, making energy conservation especially important.

The filters used in a satellite repeater are a preselection filter, filters used for the input multiplexer (IMUX), filters used for the output multiplexer (OMUX), and finally, a lowpass filter before the antenna to eliminate unwanted high-frequency harmonics generated by the power amplifier. The preselection and IMUX/OMUX filters are commonly made of waveguide technology. The new technologies developed to reduce the typical large size and weight of these filters are described in the following sections.

Microwave filters are also very important in radar applications (both civil and military). World War II and the invention of radar contributed greatly to the development of microwave filters, in particular to the development of waveguide and coaxial filters. Radiometry (the measure of electromagnetic radiation) is another application in which microwave filters are important.

In this chapter we present a bibliographic study of the most important technologies and techniques used to provide microwave filters. These technologies and techniques are divided into two main categories: those based on waveguides and those based on planar structures. This study can be used to define the technology best suited for a given application and set of requirements.

1.2 CAVITY FILTERS

Cavity filters are used for their excellent electric performance. They can provide very high quality factors (Q factors), and it is possible to design very selective filters with low insertion losses. In addition, due to the distances encountered in satellite communications, the terrestrial transmission filters must be able to handle large power levels. Cavity filters can support these high power requirements.

There are several types of cavity filters: evanescent-mode waveguide filters, coupled cavity filters, dielectric resonator filters, and E-plane filters. These are described in some detail below.

1.2.1 Evanescent-Mode Waveguide Filters

From theory, there are a multitude of propagation modes that satisfy Maxwell's equations in a waveguide. However, some of these modes will be greatly attenuated after a short distance inside the waveguide, depending on the frequency. This will happen for frequencies that are below the associated cutoff frequency of a propagation mode. For monomode filters, only one mode of propagation is used. The simplest solution is to use the mode that has the lowest cutoff frequency (the fundamental mode). Therefore, one must choose a waveguide such that the cutoff frequency of the fundamental mode is below the frequency band of interest (passband of the filter) and the cutoff frequency of the second mode is above this band. Unfortunately, the dimensions of a waveguide are inversely proportional to the cutoff frequency, and the fundamental mode has the lowest cutoff frequency.

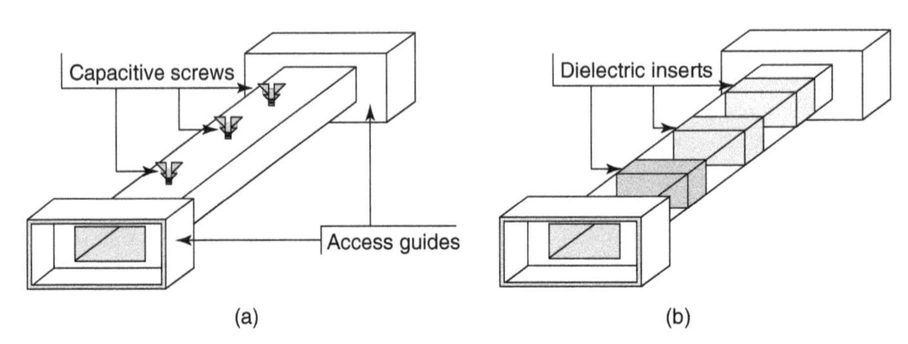

Figure 1.3 Evanescent-mode waveguide filters: (a) with capacitive screws; (b) with dielectric inserts.

In evanescent-mode waveguide filters, the dimensions of the waveguide can be reduced by relaxing the cutoff frequency requirements. In this case, all the modes of propagation can be evanescent and their amplitude decreases exponentially with the distance traveled inside the waveguide.

To obtain filtering characteristics, capacitive elements are placed at very specific locations inside an evanescent-mode waveguide. In practice, the filtering characteristics are defined only for the fundamental mode, since it is assumed that all the higher-order modes will vanish or be greatly attenuated between these capacitive elements. Inside the capacitive elements, only the fundamental mode is propagating and therefore is the only mode to "survive" the entire structure.

One of the first evanescent-mode waveguide filter designs was proposed by Craven and Mok [1] in 1971. The filter is made of an evanescent-mode waveguide and capacitive screws as shown in Figure 1.3(a). A set of formulas provides the order of magnitude needed for the capacitive elements from which the size of the screws can be defined. It also provides the positions at which the screws must be inserted. Unfortunately, it does not provide the depth at which the screws must be inserted. This must be done using a manual tuning technique once the filter has been built. The method provides good results for filters with relative bandwidths up to 20%. In 1977 and 1983, Snyder [2,3] extended the technique to design filters with larger bandwidths (up to an octave) using an admittance inverter representation of the structure which included the frequency dependence of the inverters.

In 1984, Shih and Gray [4] proposed replacing the capacitive screws with dielectric inserts, as shown in Figure 1.3(b). This type of evanescent-mode waveguide filter has been studied exhaustively at the IMS at the University of Bordeaux since the late 1980s, and new and more exact design methods have been defined. The most recent realizations of such filters are given in the Ph.D. work of Lecouvé [5] and Boutheiller [6]. The dielectric permittivity used is small (Teflon, $\varepsilon_r \approx 2.2$) but allows the fundamental mode to be propagating in the dielectric insert. The structure can be thought of a series of resonators coupled through evanescent-mode waveguide sections. The design method allows

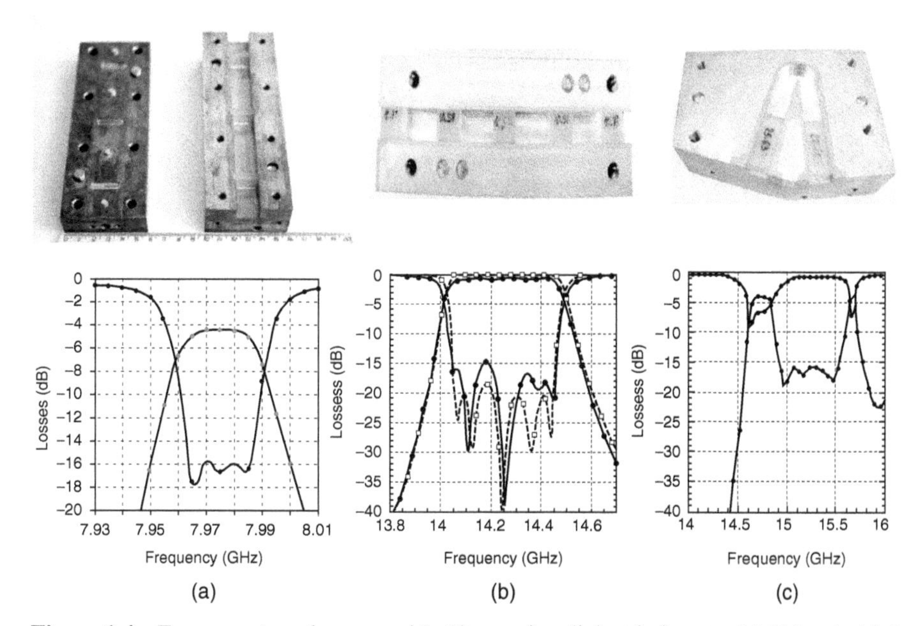

Figure 1.4 Evanescent-mode waveguide filters using dielectric inserts: (a) X band, third order; (b) Ku band, fifth order; (c) folded waveguide, third order.

defining the exact positions and length of the dielectric inserts, and no manual tuning is necessary.

Figure 1.4 shows several realizations and measured responses of evanescent-mode waveguide filters using dielectric inserts. Realizations (a) and (b) use straight waveguides to provide specified Chebyshev responses. Realization (c) uses a waveguide that includes a bend with a given angle of curvature. This method can be used to reduce the form factor of the filter. Due to the extreme complexity of defining the S parameters of all the modes in such a bend, the design technique is still in development, as can be seen in the actual response of the filter. One also observes that the response has three reflection zeros (consistent with a third-order Chebyshev function) but that only two dielectric inserts are used. This means that a bend can produce an additional reflection zero to increase the order of the filter. This opens the door for a new category of evanescent-mode waveguide filters.

1.2.2 Coupled Cavity Filters

Coupled cavity filters rely on the resonance of several degenerated modes inside the same cavity (multimode cavities). This concept was proposed by Lin in 1951, but it was not until the late 1960s that the first realizations appeared [8,9]. The first and most common implementation of these filters uses bimodal circular cavities. Figure 1.5(a) shows an eighth-order bimodal circular cavity filter centered at

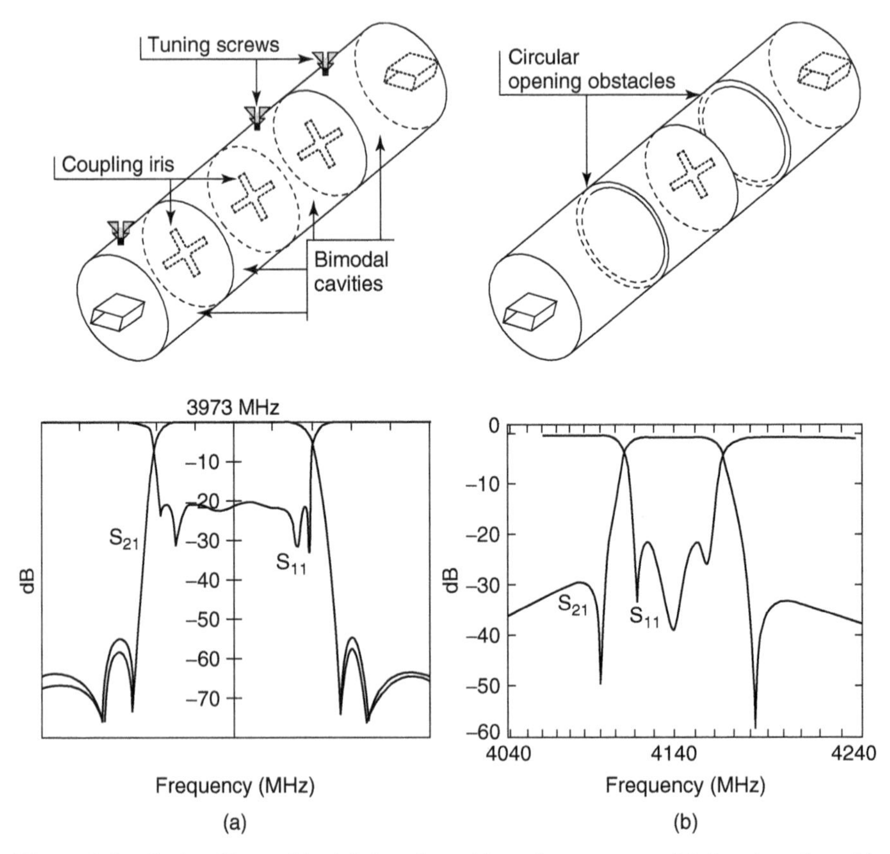

Figure 1.5 Cavity filters: (a) eighth order with tuning screws; (b) fourth order with circular opening obstacles.

3973 MHz. The tuning screws are used to fine tune the coupling coefficients and the overall response of the filter. Recent studies on this type of filter showed that the tuning screws could be replaced by circular [10] or rectangular [11] opening obstacles. Figure 1.5(b) shows the case of a fourth-order filter centered at 4140 MHz where circular opening obstacles are used. These filters are used for very narrowband applications: for example, for multiplexers. They can provide very high Q factors for the modes used, good precision cavities are easy to fabricate, and the tuning ranges are large and nearly independent.

The second implementation of these filters uses rectangular cavities. In this case, coupling between two cavities can be achieved using a small evanescent-mode waveguide section, as shown in Figure 1.6. A wide range of realizable coupling factors can be obtained by changing the dimensions of the evanescent-mode waveguide [12]. In addition, the coupling tuning screws can be replaced by small cuts in the corners of the rectangular waveguide. Figure 1.6 shows a fourth-order filter centered at 8.5 GHz using bimodal

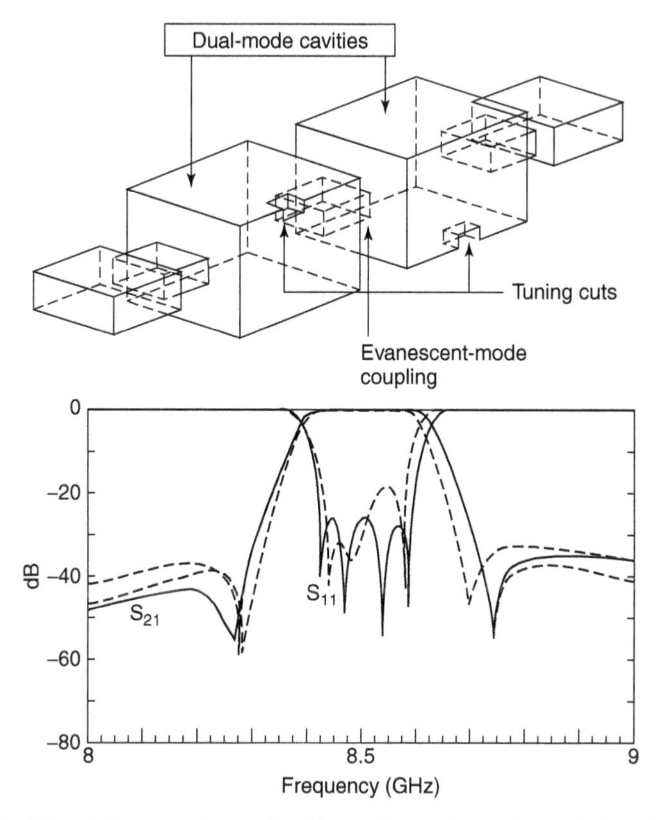

Figure 1.6 Bimodal rectangular cavity filter with small cuts instead of tuning screws.

rectangular cavities with small cuts in the waveguide instead of tuning screws [13]. As was the case for circular cavity filters, there is a wide range of rectangular cavity filters. For instance, bimodal rectangular cavity filters that do not rely on degenerated modes have been designed and tested at the IMS at the University of Bordeaux. The dimensions of the cavity must be increased so that the fundamental mode and another mode can have the same resonant frequency [14–16]. The height of the structure can be kept constant by using inductive-type irises. This technique allows each bimodal cavity to produce two poles and one transmission zero. Figure 1.7 shows three filters designed using this technique. The filter in Figure 1.7(a) uses one bimodal cavity and is a second-order filter with one transmission zero on the right side of the passband. The filter in Figure 1.7(b) uses one bimodal cavity and one monomodal cavity. The filter is therefore third-order and has one transmission zero on the left of the passband. The filter in Figure 1.7(c) uses two bimodal cavities. It thus has four poles and two transmission zeros, one on each side of the passband.

Two CAD packages for the design and simulation of such filters were developed at the IMS by Boutheiller [6]: (1) COFRIM [conception par optimization

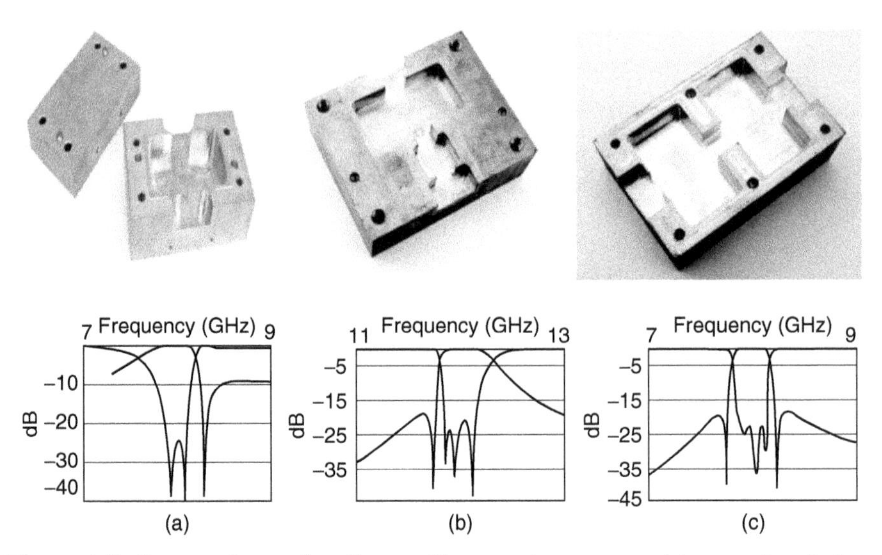

Figure 1.7 Rectangular cavity filters without tuning screws: (a) second order/one bimodal cavity; (b) third order/one bimodal cavity/one monomodal cavity; (c) fourth order/two bimodal cavities.

de filtres rectangulaires à iris multisections (design using optimization of multisection iris rectangular waveguide filters)] and (2) SOFRIM [simulation et optimisation de filtres rectangulaires à iris multisections (simulation and optimization of multisection iris rectangular waveguide filters)]. These CAD packages made it possible to realize rectangular multimodal cavity filters that had transmission zeros with the smallest number of cavities by exploiting the coupling between different resonant modes within the cavities. For example, Figure 1.8(a) shows that a fourth-order filter with two transmission zeros can be obtained using only one cavity. Figure 1.8(b) shows that a seventh-order filter with two transmission zeros can be built using only two cavities.

The idea of trimodal circular cavity filters was introduced by Atia and Williams in 1971 [17]. Two TE_{111} modes polarized orthogonally are used with the TM_{010} mode. However, a new type of iris was needed to produce the required couplings independently. It was introduced by Tang and Chaudhuri in [18] in 1984. The structure of a sixth-order filter made of two trimodal circular cavities is shown in Figure 1.9(a). Later, a trimodal rectangular cavity filter was presented by Lastoria *et al.* [19]. The rectangular cavity filters use the TE_{10}, TM_{11}, and TE_{01} modes. The coupling between cavities is obtained using two rectangular metallic strips placed along the edges of the cavity as shown in Figure 1.9(b).

A quadruple-mode cavity would require two orthogonal modes that could be used twice using different polarizations. In 1987, Bonetti and Williams [20] used the TE_{113} and TM_{110} resonant modes. However, this structure is sensitive to slight changes in dimensions and the filter performance is difficult to maintain over a wide range of temperatures.

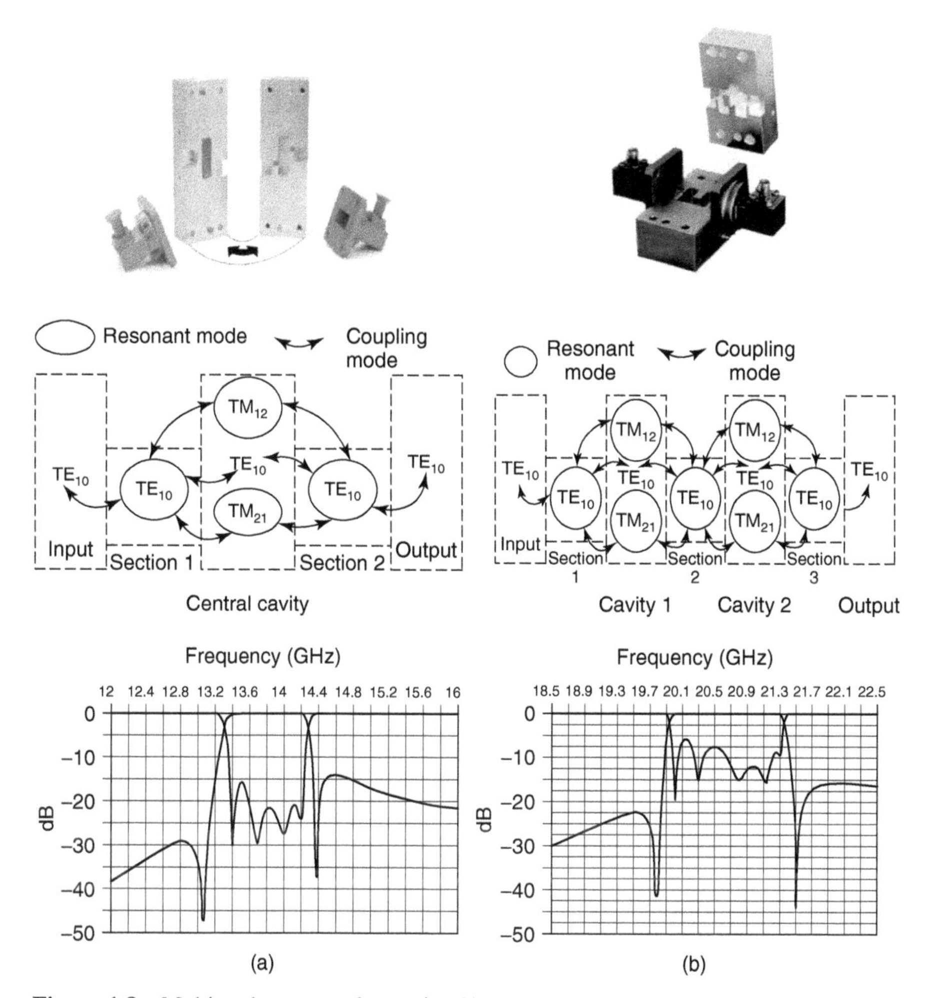

Figure 1.8 Multimode rectangular cavity filters: (a) fourth order/two zeros; (b) seventh order/two zeros.

A few cavity filters have been reported that use more than four modes. In 1992, Sheng-Li and Wei-Gan [21] were able to realize a spherical cavity that had five resonant modes and produced two transmission zeros. However, the different couplings and overall response of the filter must be tuned manually using screws. A few years earlier, in 1990, Bonetti and Williams [22] achieved a similar result with a rectangular cavity. The structure had six resonant modes (TE_{102}, TE_{201}, TE_{021}, TE_{012}, TM_{120}, and TM_{210}).

1.2.3 Dielectric Resonator Filters

The use of dielectric resonator filters appears in the late 1960s. They are based on the use of cylindrical [23] or rectangular [24] resonators. The dielectric resonators

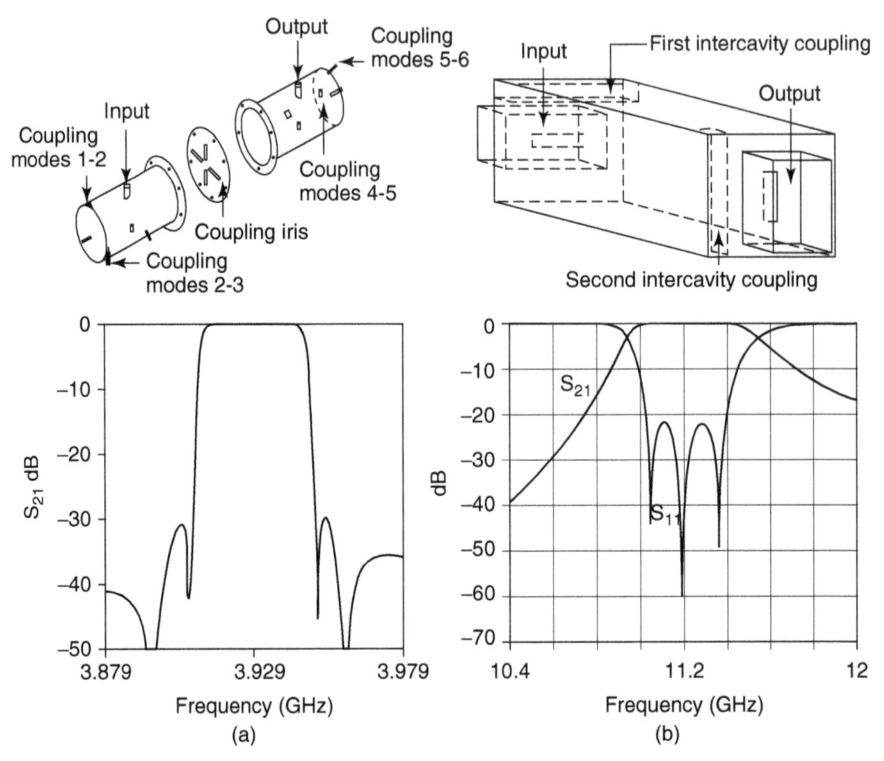

Figure 1.9 Trimodal cavity filters: (a) sixth order/triple-mode cylindrical cavity; (b) triple-mode rectangular cavity.

are placed inside a metallic cavity to avoid losses by radiation. At this stage they can be excited following a transverse electric TE_{0n} mode, a transverse magnetic TM_{0m} mode, or following a hybrid TEM_{nm} mode. The fundamental mode depends on the diameter/resonator height ratio. Figure 1.10 shows possible dielectric resonator filter layouts. In Figure 1.10(a) the dielectric resonators are placed axially; in Figure 1.10(b) the resonators are placed laterally inside the metallic cavity. The cavity can be circular or rectangular.

The required coupling between resonators can be determined using an approximate method based on magnetic moments proposed by Cohn [25]. The design techniques presented for bimodal cavity filters can be used to design dielectric resonator filters. Tuning screws are often necessary. Coupling between cavities is achieved using a traditional iris [26] as shown in Figure 1.11(a), or by using a section of waveguide [27] as shown in Figure 1.11(b).

A serious drawback of a dielectric resonator is its low thermal conductivity. This can produce large temperature variations within the resonators, which in turn degrades the response of the filter. A solution to this problem consists in inserting metallic conducting planes inside the dielectric to increase its thermal conductivity [28]. Another solution is to use resonators with more complex geometry,

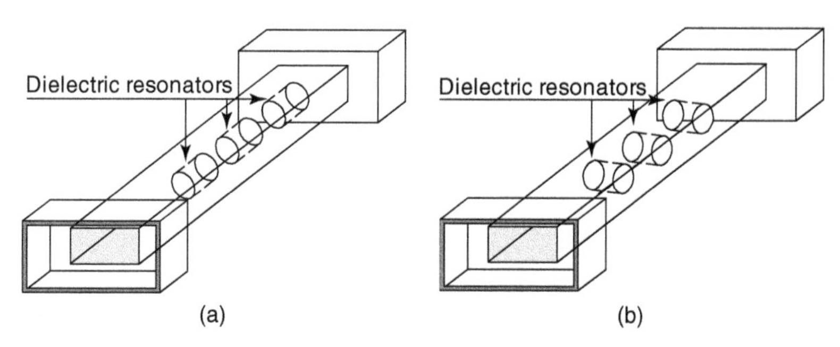

Figure 1.10 Dielectric resonator filters: (a) axially positioned resonators; (b) laterally positioned resonators.

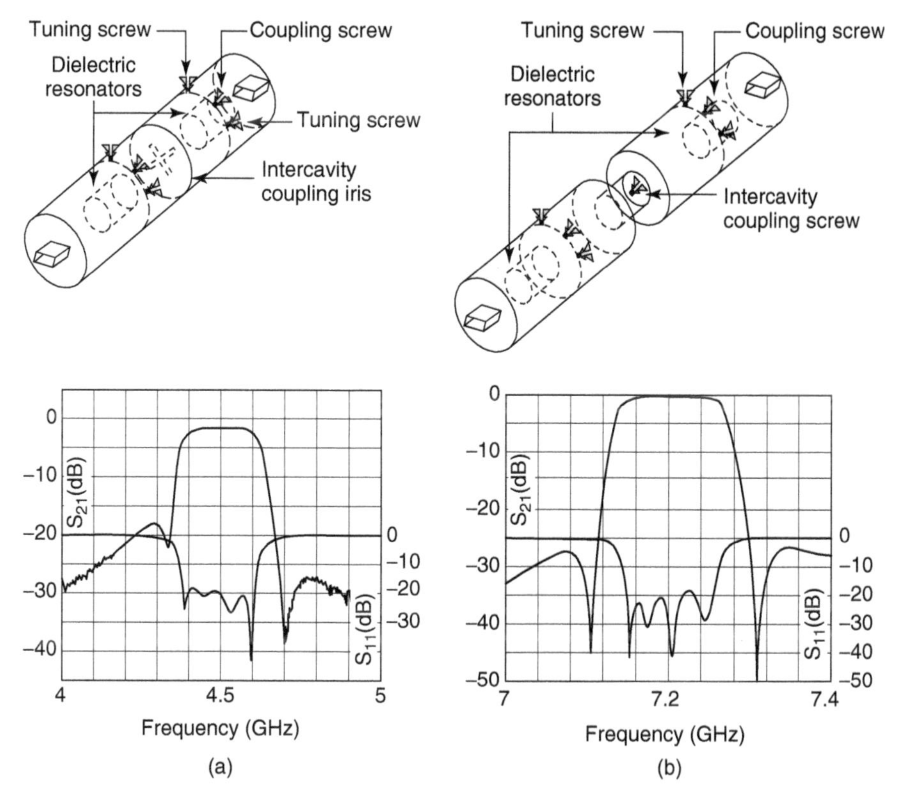

Figure 1.11 Dielectric resonator filters: (a) tuning screw coupling; (b) waveguide coupling.

such as octagonal resonators [29]. Another problem comes from the difficulty in maintaining the dielectric resonator in the correct position, especially for space applications, where vibrations can be severe.

The input/output of a filter can be achieved using waveguides or magnetic or electric probes to reduce insertion losses. Excitation through a microstrip line can ease integration of the filter with a planar environment.

1.2.4 E-Plane Filters

E-plane filters consist of dielectric plates with partially metallized surfaces placed in the plane of the waveguide electric field as shown in Figure 1.12(a). To realize bandpass filters with the desired filtering characteristics, metallization is used to produce half-wavelength resonators that are coupled between themselves as shown in Figure 1.12(b). To increase the Q factor of the filter, the dielectric substrate can be eliminated, and the filter is then made of a metallic sheet placed between the two halves of the waveguide, as shown in Figure 1.13(a). These filters, called *filters with metallic inserts*, should not be mistaken for finline filters. *Finline filters* have a dielectric substrate (see Table 1.1). A drawback, however, is that filters with metallic inserts are more sensitive than finline filters to fabrication tolerances. Filters with metallic inserts can be designed using a method based on equivalent-circuit representations [30,31] or using a method based on optimization and electromagnetic characterization of the filter where the structure is divided into homogeneous waveguide sections [32].

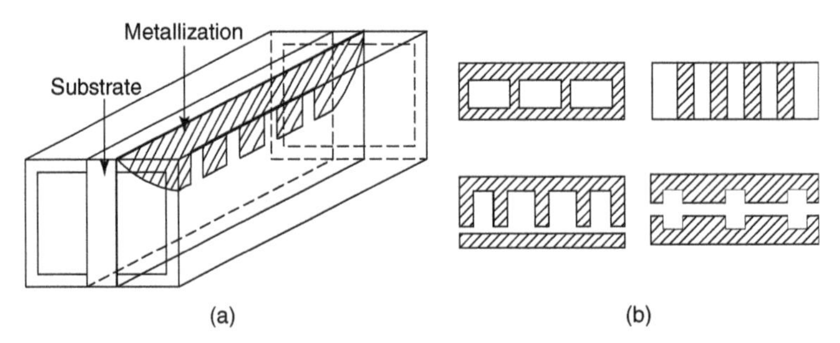

Figure 1.12 E-plane filters: (a) filter structure; (b) metallization types.

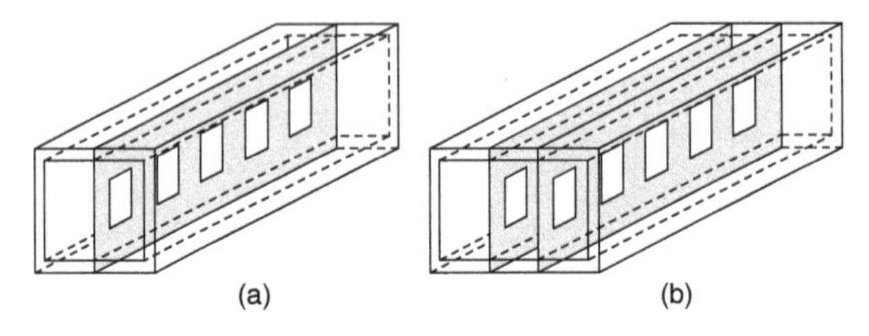

Figure 1.13 E-plane filters with metallic inserts: (a) one insert; (b) two parallel inserts.

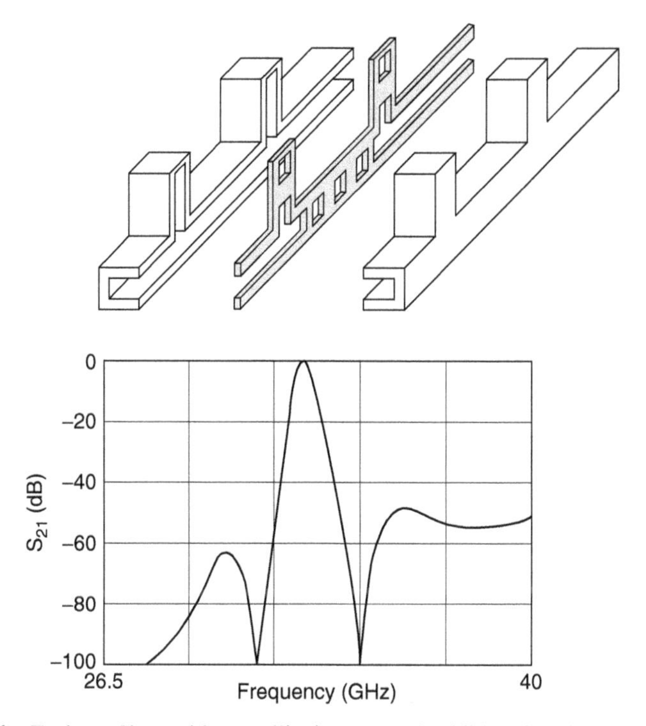

Figure 1.14 E-plane filter with metallic inserts and additional stubs to produce two zeros.

The stopband rejection of the filter can be improved by placing two metallic inserts in parallel as shown in Figure 1.13(b). By reducing the distance between the inserts and the waveguide walls, the fields coupling the two resonators become evanescent and provide inductive coupling. The selectivity of the filter can also be improved by introducing additional waveguide sections that act as stubs, as shown in Figure 1.14. This configuration can produce transmission zeros [33]. Figure 1.14 shows an example of an E-plane filter with a metallic insert that has a transmission zero on each side of the passband.

1.3 PLANAR FILTERS

Planar filters have seen important developments throughout the years in terms of topologies, design methods, and realization technologies. Planar filters have several advantages. They are small in size, are capable of being fully integrated with surrounding electronic circuits, their production costs are low, and they can easily be reproduced. Their main drawback is a poor unloaded Q factor that results in poor selectivity and serious insertion losses. To counter this problem, new technologies (i.e., superconductivity, micromachinery, multilayer technology, etc.) have been proposed to improve the electrical performances.

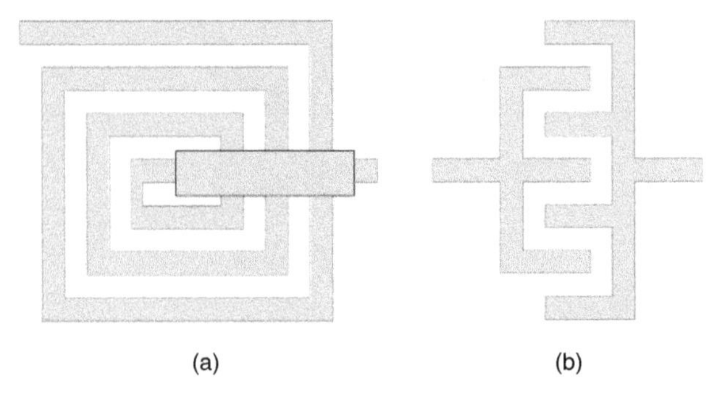

Figure 1.15 Semi-lumped elements: (a) inductor; (b) capacitor.

1.3.1 Semi-lumped Filters

Semi-lumped filters are realized using inductors and capacitors made with integrated-circuit techniques. For example, an inductor can be obtained using a spiral line, as shown in Figure 1.15(a), and a capacitor can be obtained using two intertwined lines, as shown in Figure 1.15(b). These filters have high levels of integration but have poor Q factors and high losses [34]. The losses can be compensated using amplifying devices for active filter applications. Fabrication quality and inaccurate semi-lumped element models can lead to undesired frequency shifts of the filter response. Electrical tuning techniques to control the frequency are often needed to improve the filter response. These filters are used for applications up to 5 GHz. Above 5 GHz, it is preferable to use other technologies that can provide better Q factors.

1.3.2 Planar Transmission Line Filters

Coupled Line Filters These filters are made of metallic lines of length $\lambda_g/2$ or $\lambda_g/4$, where λ_g is the wavelength placed on a dielectric substrate. They are particularly attractive for applications where the wavelengths are small, such as high-frequency or microwave applications. The width of the line defines the characteristic impedance of that section.

Coupled line filters are based on line resonators that are coupled and terminated by either open or short circuits. The filter response depends on the number of resonators (order of the filter), on the distance between the line resonators (gaps), and on the properties of the substrate. The most common topologies are parallel coupling, where the lines are placed side by side over a length of $\lambda_g/4$ as shown in Figure 1.16(a), and end coupling, where the lines are placed one after the other as in Figure 1.16(b).

Parallel coupling is stronger than end coupling and can be used to provide filters with wider passbands. A parallel-coupled filter can have half the length of an end-coupled filter. In this type of filter the spurious response (undesirable second passband) occurs at three times the central frequency of the filter. The synthesis of these filters was introduced in the late 1950s by Cohn [35,36] but

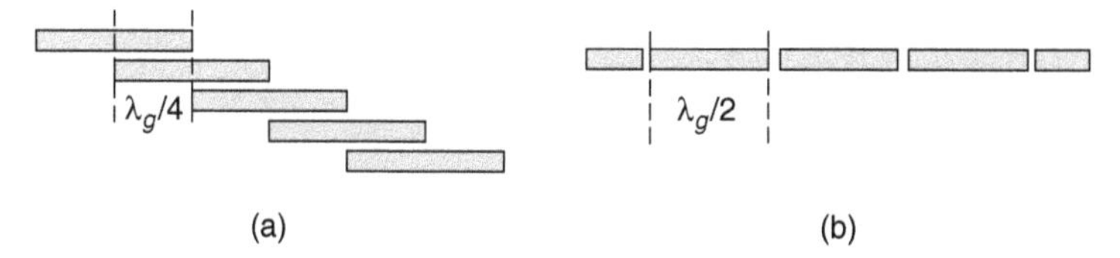

Figure 1.16 Coupling topologies: (a) parallel coupling; (b) end coupling.

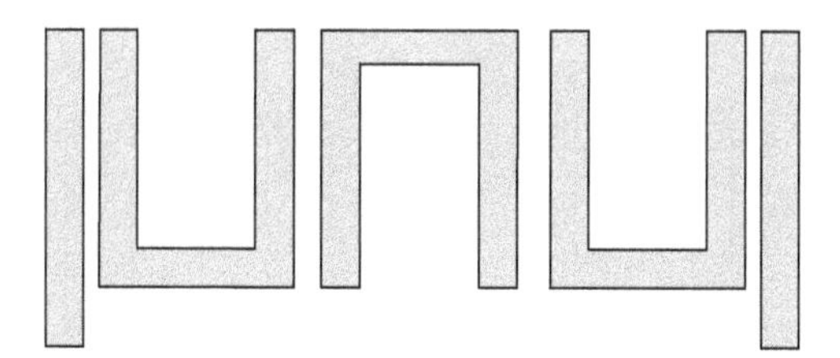

Figure 1.17 Structure of a hairpin filter.

is only applicable to filters with relative bandwidths up to 30%. Another line resonator–based filter is the hairpin filter shown in Figure 1.17 [37,38]. The coupling used in these filters is of the parallel type.

Interdigital and Pseudo-interdigital Filters As a follow-up to his work on parallel-coupled $\lambda_g/2$ resonator filters, Matthaei developed a design method for interdigital filters in 1962 [39]. The exact theory for the synthesis of these filters was later provided by Wenzel [40]. Interdigital filters are made of parallel conductors placed transversally between two ground planes, as shown in Figure 1.18. One side of the conductor is connected to the ground and the other side is left open. These resonators all have the same length ($\lambda_g/4$). There are two main possibilities for connecting the filter to the rest of the system. In the first case, the access is achieved in open-circuit fashion, and these conductors are included in the number of resonators of the filter, as shown in Figure 1.18(c). In the second case, the access is done in short-circuit fashion as shown in Figure 1.18(d). The second case is used to obtain wideband filters (relative bandwidths above 30%).

Interdigital filters are compact and easy to fabricate. In practice, the gaps between resonators are relatively large, so these filters are not very sensitive to fabrication tolerances. In addition, these filters can be fine tuned using external variable capacitors. However, the insertion losses are important, particularly for narrowband filters. The unloaded Q factors are low, and interdigital filters have limited power handling.

A variation of the interdigital filter, the combline filter (Figure 1.19), was proposed by Matthei in 1963 [41]. The structure is nearly identical to the interdigital filter of Figure 1.18, except that the line resonators are all shorted on one side and loaded with a variable capacitor on the other. This can be used to realize electronically tunable filters [42]. In practice, the length of the lines is

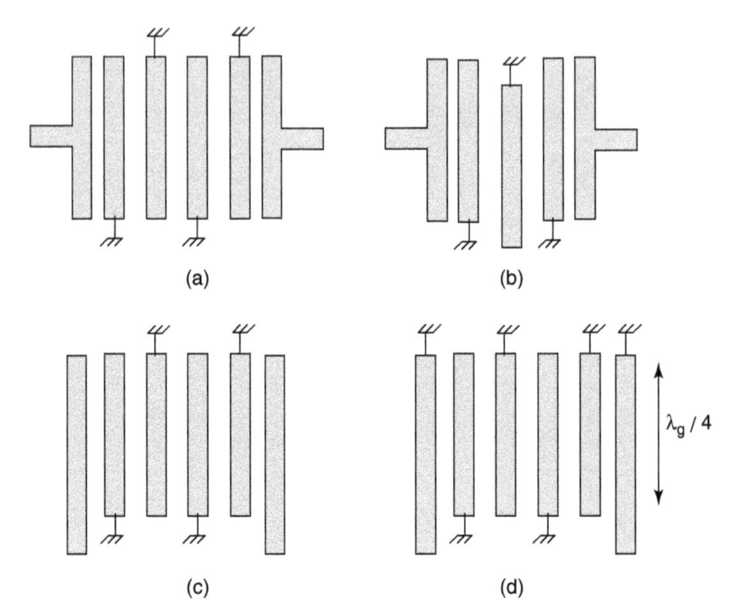

Figure 1.18 Interdigital filters: (a) interdigital; (b) pseudo-interdigital; (c) open-circuit access; (d) short-circuit access.

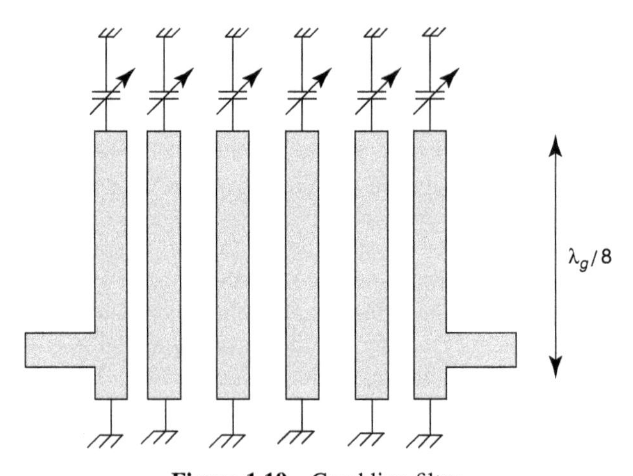

Figure 1.19 Combline filter.

about $\lambda_g/8$, so combline filters are more compact than pure interdigital filters. The spurious response occurs at four times the central frequency and represents an improvement over interdigital filters.

Stub Filters Stub filters are made of $\lambda_g/2$ or $\lambda_g/4$ lines and a number of stubs, as shown in Figure 1.20. This topology can produce compact structures with small insertion losses. The stubs can be used to produce transmission zeros in order

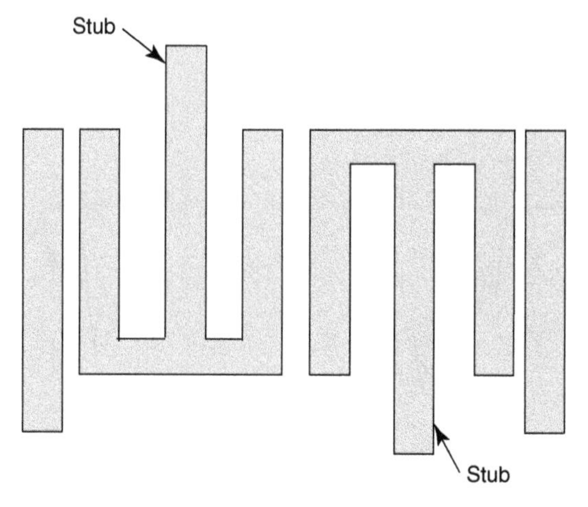

Figure 1.20 Stub filters.

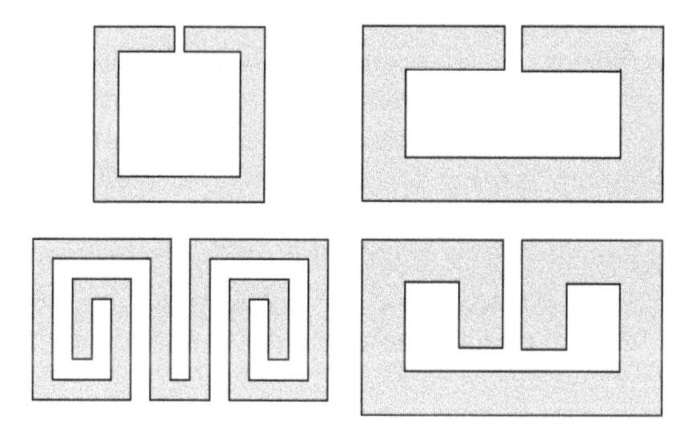

Figure 1.21 Bended $\lambda_g/2$ line resonators.

to improve stopband rejection [43]. There exist other filter topologies based on line resonators. To improve the transfer characteristics of the filter and to make them more compact, new topologies use $\lambda_g/2$ lines that are bent as shown in Figure 1.21. Figure 1.22 shows trisection filters using $\lambda_g/2$ line resonators in the shape of a rectangle. Depending on the orientation and position of the resonators, the filters can produce transmission zeros on the right or left of the passband. This type of filter has been studied primarily by Hong and Lancaster [44–46].

Regardless of the shape of the resonator, the synthesis of a filter using line resonators usually follows the same steps. At first, theoretical research is done to define the theoretical coupling coefficients associated with a lowpass filter prototype that satisfies the filtering requirements. Then experimental research using electromagnetic simulations is done to define the coupling between two resonators in terms of the gap distance that separates them. The physical filter is then constructed by finding the length of the various gaps that give the theoretical couplings.

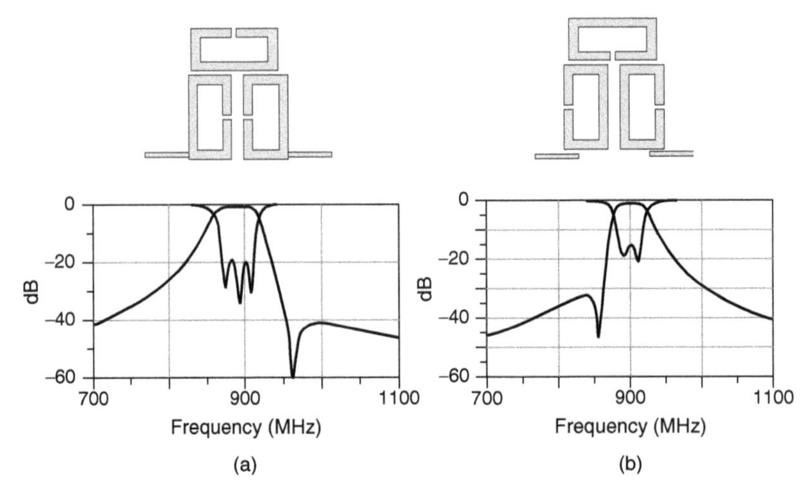

Figure 1.22 Trisection filters using rectangular $\lambda_g/2$ line resonators.

Patch Resonator Filters Patch resonators, also called *bimodal resonators*, are surface resonators of various geometries (rectangular, square, triangular, circular), as shown in Figure 1.23. These resonators are used primarily for the realization of antennas, but they are starting to be used for the realization of microwave filters. They have two resonant modes that are orthogonal. This is achieved by adding or removing small pieces to alter the symmetry of resonator so that the two resonant modes can be decoupled. This can reduce by half the number of resonators needed for a given filter order compared to a traditional line resonator

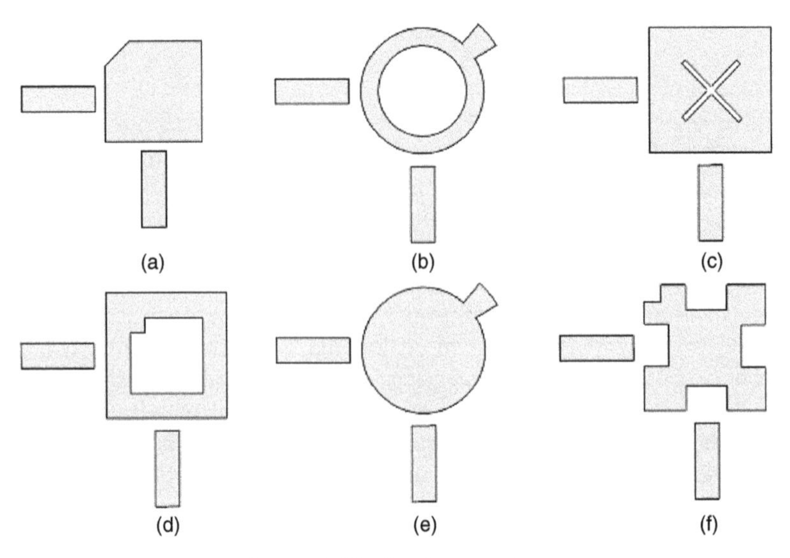

Figure 1.23 Patch resonators.

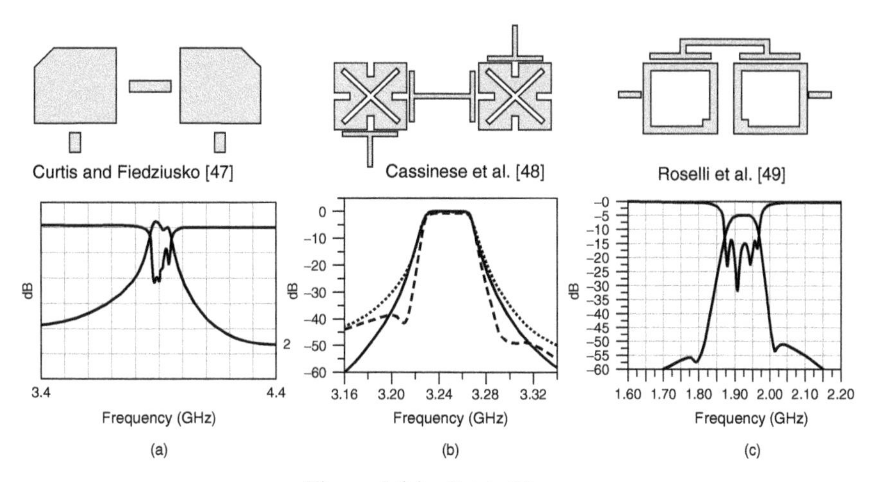

Figure 1.24 Patch filters.

filter. In addition, bimodal resonators have good power handling. With patch resonators it is possible to realize filters with elliptic responses using a small number of resonators. For instance, a fourth-order elliptic function with two transmission zeros can be obtained using only two patch resonators. Several patch resonator filters and their frequency responses are shown in Figure 1.24.

1.4 PLANAR FILTER TECHNOLOGY

In this section the most important technologies used for the realization of planar filters are reviewed. These are microstrip, coplanar, suspended substrate stripline, and multilayer technologies.

1.4.1 Microstrip Technology

Microstrip technology consists of a thin conducting strip placed on top of a dielectric substrate. The bottom side of the substrate is covered entirely by metal and serves as the ground plane, as shown in Figure 1.25. The substrate is characterized by its dielectric permittivity and thickness. It serves as a mechanical support

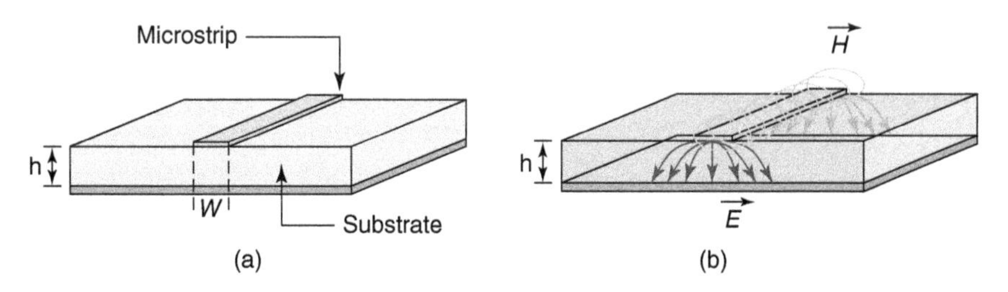

Figure 1.25 Microstrip technology.

TABLE 1.1 Microstrip Technology Variations

Technology	Basic Configuration	Variations		
Microstrip		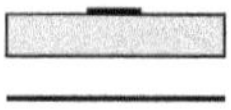 Suspended	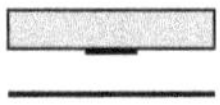 Inverted	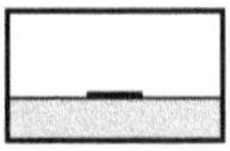 In box
Stripline	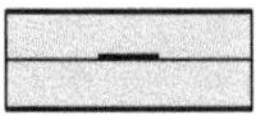	Stripline with two conductors	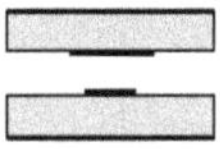	
Stripline		 Antipodal		 Bilateral
Finline	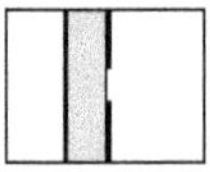	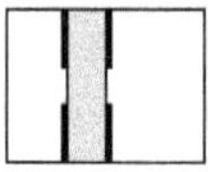 Bilateral	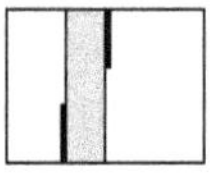 Antipodal	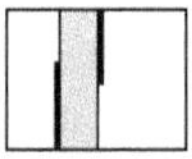 Antipodal and strong coupling

and as a medium for electromagnetic field propagation. In 1975, Hammerstad [50] proposed the first equations to compute the characteristic impedance of a microstrip line. It is proportional to the ratio of the strip width W over the substrate thickness h. Since then, additional equations have been proposed to account for the strong dispersion of microstrip lines at millimeter wavelengths. Other technologies that use conducting strips, substrate, and metal ground planes are summarized in Table 1.1.

1.4.2 Coplanar Technology

In the case of coplanar technology, additional ground planes surround the conducting strip on the top of the substrate as shown in Figure 1.26. This avoids having to drill through the substrate to bring up the ground, a technique that introduces parasitic effects. Another advantage of coplanar technology is the ease of integration with other system elements, such as those made in MMIC technology [51]. A drawback, however, is that coplanar technology suffers from greater radiation compared to microstrip, as can be seen in Figures 1.25(b) and 1.26(b).

1.4.3 Suspended Substrate Stripline Technology

Suspended substrate appeared in the 1960s as a variation of traditional microstrip technology [64,65]. A conducting strip is placed on a thin substrate. The strip and substrate are surrounded by air and placed inside a metallic enclosure as

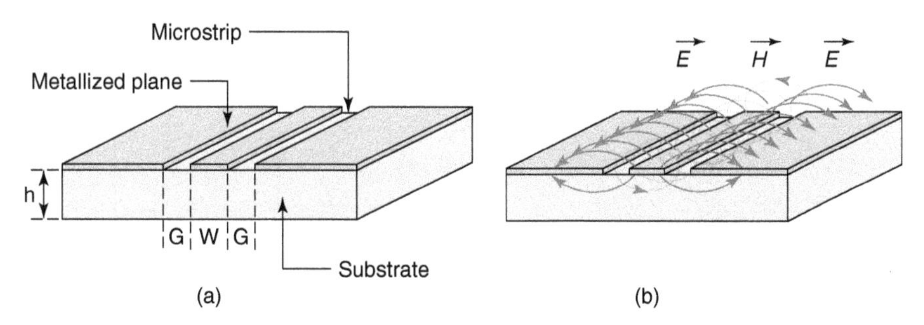

Figure 1.26 Coplanar technology.

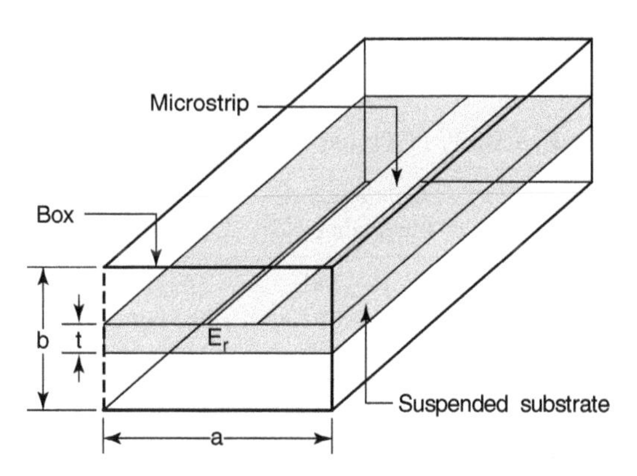

Figure 1.27 Suspended substrate technology.

shown in Figure 1.27. From the front, the substrate has the appearance of being suspended in air. This technique produces a significant improvement in the electric performance of filters compared to traditional microstrip realizations since electromagnetic field propagation is achieved mainly through the air, and there is little radiation due to the metallic enclosure. This also increases the unloaded Q factor of the resonators. The enclosure must, however, be designed carefully to avoid TEM modes that resonate at frequencies in the vicinity of the passband of the filter and can disrupt the filter propagation characteristics.

Figure 1.28 shows a 10-GHz end-coupled line filter that uses suspended substrate technology. This example uses strips on both sides of the substrate [66]. This technology is particularly interesting and is explained further in Chapter 3. Table 1.2 provides a comparative study of the different planar technologies in terms of typical operating frequencies, achievable characteristic impedances, dimensions, losses, and power-handling capabilities.

1.4.4 Multilayer Technology

Multilayer or LTCC (low-temperature cofired ceramic) technology appeared in the early 1990s in answer to cost, performance, and complexity requirements for

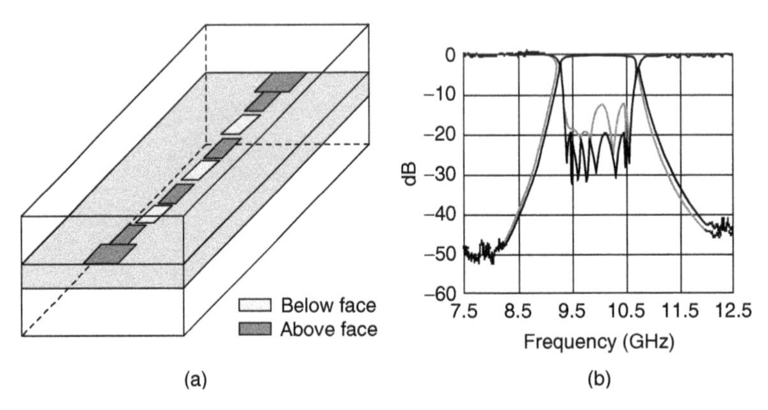

Figure 1.28 End-coupled line filter using suspended substrate technology.

TABLE 1.2 Comparative Study of Planar Technologies

Transmission Line Type	Frequency (GHz)	Characteristic Impedance (Ω)	Dimensions	Losses	Power Handling
Microstrip	<110	10–100	Small	High	Low
Stripline	<60	20–150	Medium	Low	Low
Suspended stripline	<220	20–150	Medium	Low	Low
Finline	<220	20–400	Medium	Medium	Low
Slotline	<110	60–200	Small	High	Low
Inverted microstrip	<110	25–130	Small	Medium	Low
Coplanar	<110	40–150	Small	High	Low

new applications at millimeter wavelengths. This technology is based on piling up several thin layers of dielectric and conducting materials. Lines at different layers can be connected using troughs built during the fabrication process. The pile of thin layers is then baked at 850°C so that a compact and homogeneous structure is obtained. Figure 1.29 shows a patch filter realized in LTCC technology. The passband is centered at 29 GHz, but there is about 3 dB of insertion loss [52].

With this technology it is, however, possible to integrate directly into the substrate passive elements, such as resistors, capacitors, and even inductors. This technology provides very compact designs. The height dimension provides more flexibility as to where to place the connections to the rest of the system (e.g., left, right, above, below). Unfortunately, this technology does not provide any significant improvement in losses, one of the main problems of planar filters.

1.5 ACTIVE FILTERS

To compensate for the important substrate losses in silicon MMICs, active filters can be used. Active recursive filters on silicon are based on differential circuits.

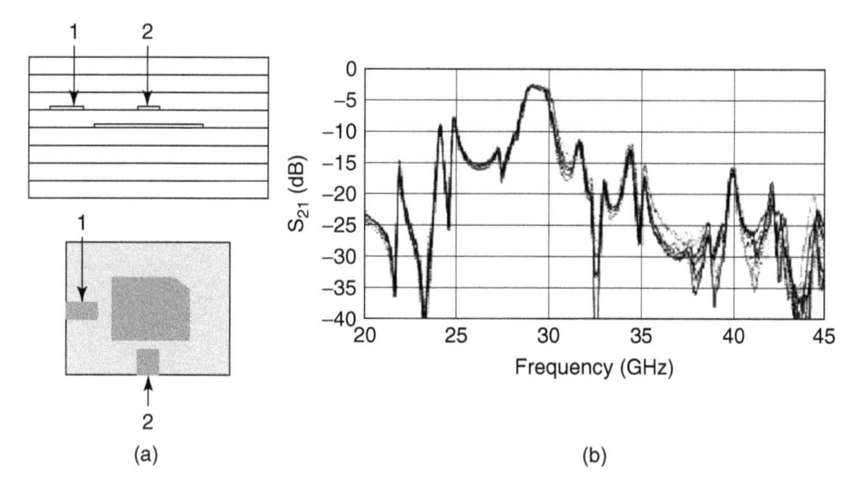

Figure 1.29 Multilayer LTCC filter.

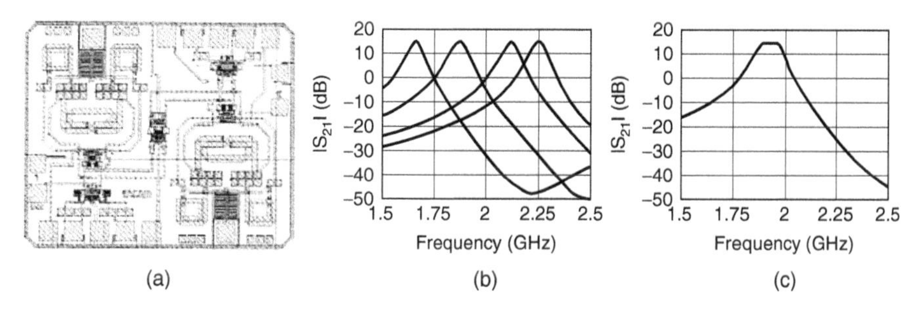

Figure 1.30 Recursive active filter.

These filters have been retained due to their immunity against electromagnetic radiations and supply and ground noises, properties that have made their success in low-frequency systems. Figure 1.30 shows an example of a differentiable tunable recursive 2-GHz filter based on the cascade of two elementary recursive stages [53]. The main drawbacks of active filters are the power consumption, the introduction of noise, and problems related to stability and linearity.

1.6 SUPERCONDUCTIVITY OR HTS FILTERS

The discovery of high-temperature superconductivity (HTS) in 1986 led to the development of a new technology suitable for planar filters. These filters use thin high-temperature superconductors and can achieve narrowband responses with very small insertion losses. The small insertion losses are obtained by the superconductors, which have very small surface resistances compared to typical conductors. This also produces very high unloaded Q factors. As an example,

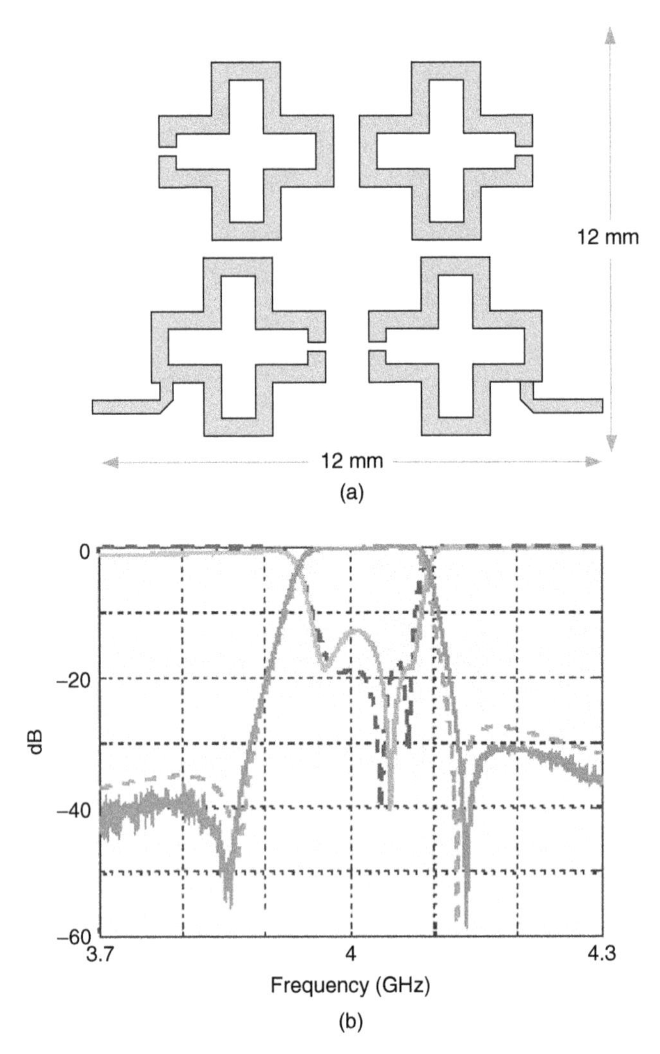

Figure 1.31 Superconductivity HTS filter.

Figure 1.31 shows a 4-GHz fourth-order elliptic filter with two transmission zeros using HTS technology [54]. Unfortunately, HTS filters require a significant cooling system, and their power-handling capability is still inferior to that of waveguide filters [55].

1.7 PERIODIC STRUCTURE FILTERS

Periodic structures such as photonic bandgaps (PBGs) have been studied since the appearance of microwave radars and were first used by Collin and Zucker to realize antennas in 1969 [56]. The main concept is that a periodic alignment of a

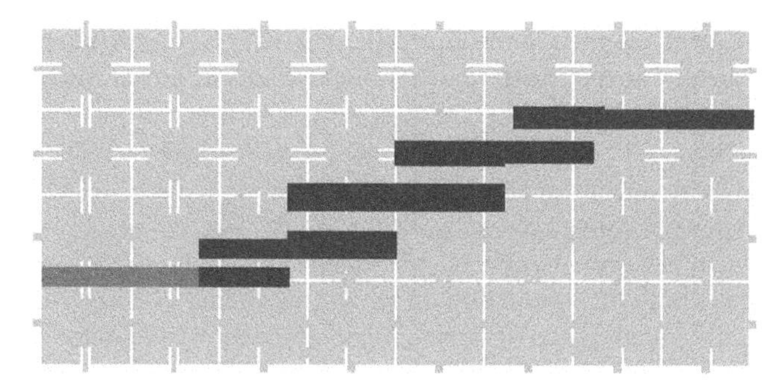

Figure 1.32 Parallel-coupled line filter with periodic pattern in the ground plane.

material could be used to block wave propagation over a given frequency band. The periodic structures can be dielectric based (holes in the substrate) or metallic based (patterns in the ground plane). Figure 1.32 shows a parallel-coupled line filter where the ground plane has periodic patterns [57]. The main drawback of this type of filter is that during fabrication the periodic pattern must be repeated throughout the entire structure to guarantee the best electric performance.

1.8 SAW FILTERS

Surface acoustic wave (SAW) filters use piezoelectric materials (quartz, ceramic, etc.) placed in thin layers on a substrate [58]. They are very popular in wireless applications such as cellular phone applications. A layer of piezoelectric material is deposited between two metallic electrodes as shown in Figure 1.33. An

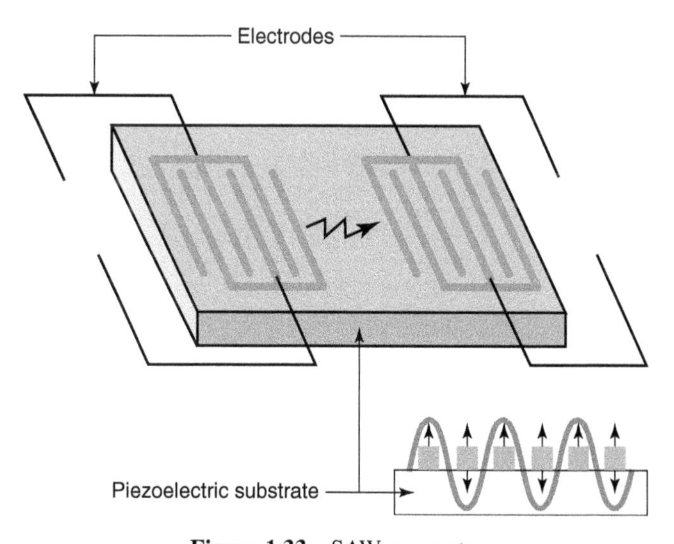

Figure 1.33 SAW concepts.

electromagnetic field applied to the electrodes generates acoustic waves that propagate on the surface of the substrate. SAW devices are compact, robust, reliable, and can be mass produced. The central frequency and filter response are achieved using photolithography, so they require very little or no subsequent tuning.

Despite all these advantages, SAW devices have some issues. The SAW component dimensions are inversely proportional to the frequency of operation. For portable applications such as cell phones, the SAW component dimensions are well within the limits of photolithography, so that manufacturing costs are low. However, with the emergence of new telecommunications standards that operate at higher frequencies (2 GHz and above), such as GPRS and UMTS, the limits of current photolithography have been reached.

In answer to these new challenges, a new technology called bulk acoustic wave (BAW), based on a three-dimensional or bulk wave resonator, has emerged [59,60]. The technique consists of inserting a piezoelectric crystal between two electrodes as shown in Figure 1.34(a). Resonance occurs when the thickness of crystal is equal to half the wavelength. The increase in operating frequency (or reduction in wavelength) is made possible by recent progress in thin-layer-deposit technology (down to a micrometer) of piezoelectric material. These techniques suggest that devices operating up to 10 GHz will be feasible. Figure 1.34(b) shows a sixth-order BAW filter that uses film bulk acoustic wave (FBAR) resonators.

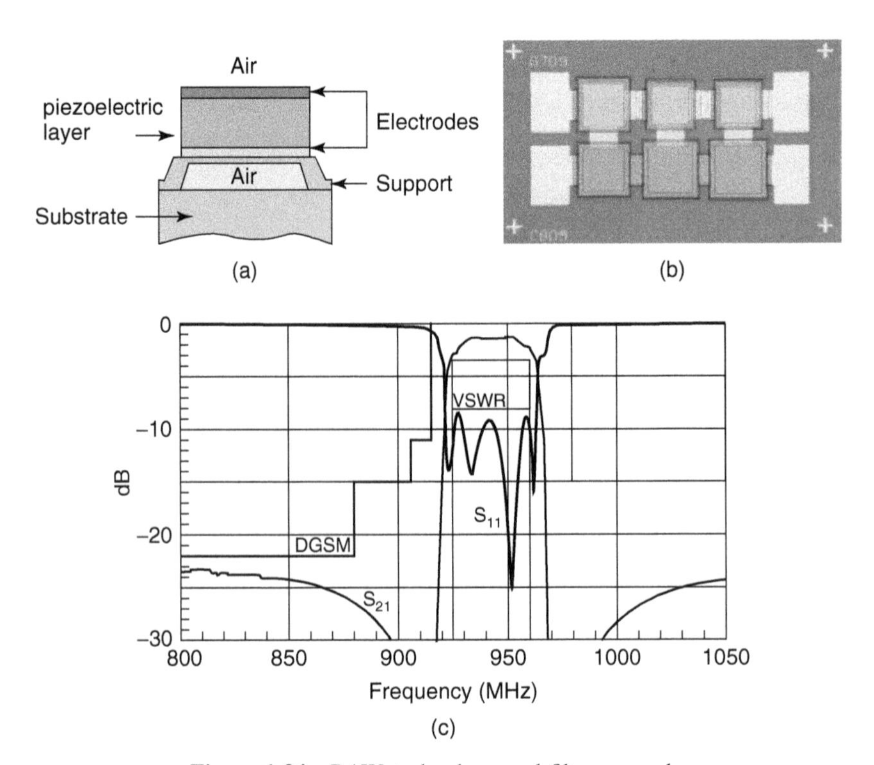

Figure 1.34 BAW technology and filter example.

The size of the filter is 1.4 × 2 mm^2. Figure 1.34(c) shows the response of the filter. It was designed for GSM applications around 900 MHz.

1.9 MICROMACHINED FILTERS

At millimeter wavelengths, microstrip and coplanar technologies are no longer adequate since dielectric and radiation losses become prohibitive. In 1991, a team from the NASA Center for Space Terahertz Technology at the University of Michigan–Ann Arbor proposed a novel type of transmission line: the micromachined line [61]. Figure 1.35(a) depicts micromachined technology. The metal conductor is deposed onto a thin suspended dielectric membrane. This reduces the dielectric losses and dispersion. The upper cavity reduces radiation losses since the circuit is now entirely shielded from the outside. In addition, this technology provides good mechanical stability and repeatability and is compatible with the MMIC elements of the filter. Figure 1.35(b) shows a micromachined interdigital

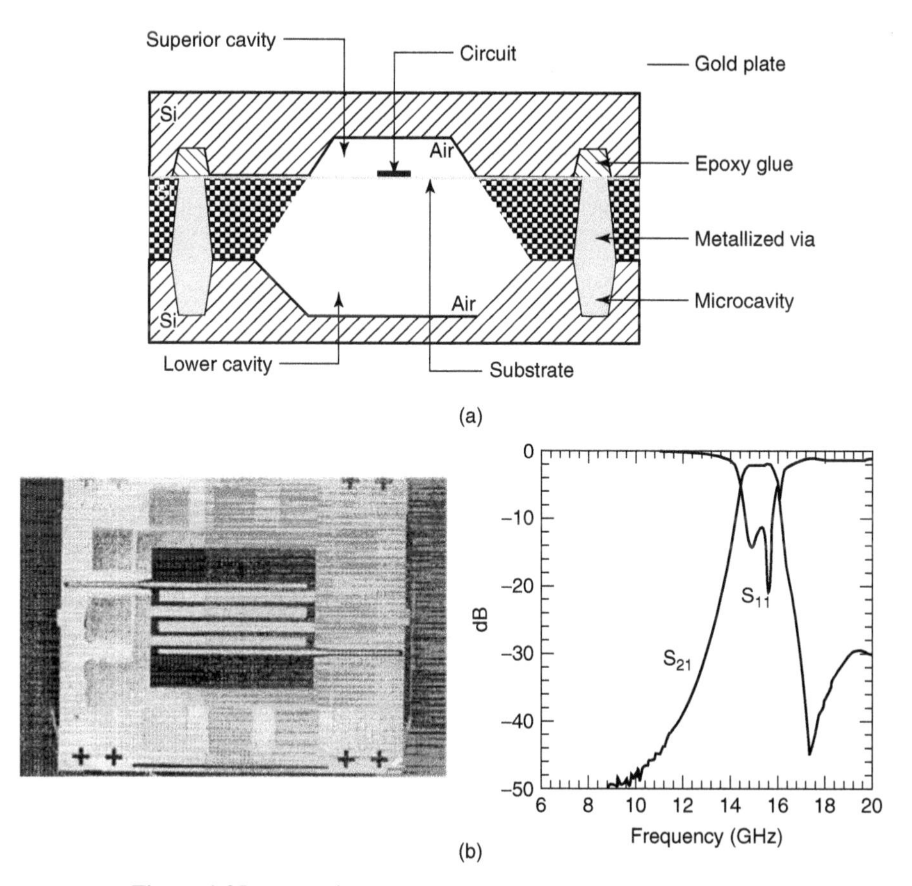

Figure 1.35 (a) Micromachined technology; (b) filter example.

filter operating at 14 GHz. Since then, micromachined silicium has been used to realize millimeter-wave devices such as filters and antennas [62,63].

1.10 SUMMARY

In this chapter we provided an overview of the various structures and technologies used for microwave filters today. The advantages and disadvantages of each technology have been given. The choice of a technology depends on several factors, such as filter response requirements, complexity and cost, size, power handling, and insertion loss levels. Table 1.3 summarizes the general properties of the two main technologies: waveguide and planar. Table 1.4 summarizes in more detail the characteristics of the technologies presented in this chapter.

TABLE 1.3 Comparative Study Between Waveguide and Planar Technologies

Technology	Advantages	Disadvantages	Other Issues
Waveguide	Small insertion losses	Fabrication tolerances	Vibrations
	High power handling	Large weight and size	
	High Q factors	Minimal integration	Fabrication technique
		Small production (tuning)	
Planar	Small weight and size	Significant insertion losses	Substrate choice
	Good integration	Small power handling	
	Large production	Poor Q factors	Metallization choice

TABLE 1.4 Summary of the Various Technologies

Technology	Frequency Range (GHz)	Size	Power Handling	Mass Production	Refs.
Evanescent mode	5–100	Large at low frequencies	High	Poor due to tuning	[1–6]
Cavity	0.03–30	Large at low frequencies	High	Poor due to tuning	[7–22]
Dielectric resonator	0.5–100	Large at low frequencies	High	Poor due to tuning	[23–29]
E plane	5–100	Large at low frequencies	High	Poor due to tuning	[30–33]
Line resonator	0.5–100	Small	Moderate	Good	[34–52]
Bimodal	0.5–100	Small	Moderate	Good	[34–52]
Active	0.001–100	Small	Moderate	Good	[53]
HTS	0.5–100	Small	Small	Good	[54,55]
Periodic	20–100	Small	Moderate	Poor	[56,57]
SAW/BAW	0.001–10	Small	Moderate	Good	[58–60]
Micromachined	20–100	Small	Moderate	Good	[61–63]
Suspended substrate	2–20	Small	High	Good	[64–66]

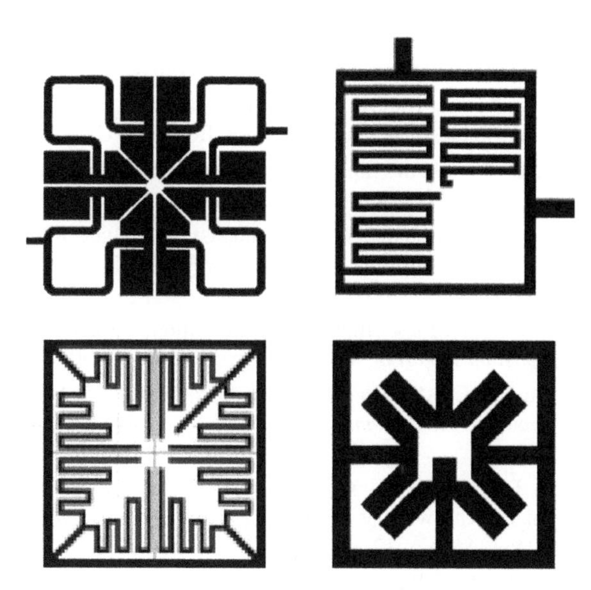

Figure 1.36 Latest planar resonator topologies.

For planar technologies, the trend is toward miniaturization of the filters by defining more complex filter topologies. Figure 1.36 shows some of the most recent planar resonators that appeared during the past few years [67–69]. The development of these resonators is based primarily on electromagnetic simulations that can quantify the coupling introduced by the various shapes and positioning. The goal is to improve the propagation characteristics while reducing the overall surface occupied by the filter. In this book, new planar resonators based on fractal iteration are shown to provide means for miniaturization while improving stopband rejection and even at times improving electrical performance such as quality factors.

REFERENCES

[1] G. F. Craven and C. K. Mok, Design of evanescent mode waveguide bandpass filters for a prescribed insertion loss characteristic, *IEEE Transactions on Microwave Theory and Techniques*, Vol. 19, No. 3, pp. 295–308, Mar. 1971.

[2] R. V. Snyder, New application of evanescent mode wave-guide to filter design, *IEEE Transactions on Microwave Theory and Techniques*, Vol. 25, No. 12, pp. 1013–1021, Dec. 1977.

[3] R. V. Snyder, Broadband waveguide filters with wide stopbands using a stepped-wall evanescent mode approach, *1983 IEEE MTT-S International Microwave Symposium Digest*, Vol. 83, No. 1, pp. 151–153, 1983.

[4] Y. C. Shih and K. G. Gray, Analysis and design of evanescent-mode waveguide dielectric resonator filters, *1984 IEEE MTT-S International Microwave Symposium Digest*, Vol. 84, No. 1, pp. 238–239, 1984.

[5] M. Lecouvé, Conception et réalisation de filtres microondes à modes évanescents à l'aide d'un algorithme génétique, thesis, Bordeaux University, Oct. 2000.

[6] N. Boutheiller, Analyse et synthèse par optimisation basée sur l'algorithme génétique de filtres en guide d'ondes rectangulaire, thesis, Bordeaux University, Sept. 2002.

[7] W. G. Lin, Microwave filters employing a single cavity excited in more than one mode, *Journal of Applied Physics*, Vol. 22, pp. 989–1001, Aug. 1951.

[8] A. E. Atia and A. E. Williams, Narrow-bandpass waveguide filters, *IEEE Transactions on Microwave Theory and Techniques*, Vol. 20, No. 4, pp. 258–265, Apr. 1972.

[9] A. E. Williams, A four-cavity elliptic waveguide filter, *1970 G-MTT International Microwave Symposium Digest of Technical Papers 70.1*, pp. 90–93, 1970.

[10] A. E. Williams and A. E. Atia, Dual-mode canonical waveguide Filters, *IEEE Transactions on Microwave Theory and Techniques*, Vol. 5, No. 12, pp. 1021–1026, Dec. 1977.

[11] L. Accatino, G. Bertin, and M. Mongiardo, A four-pole dual mode elliptic filter realized in circular cavity without screws, *IEEE Transactions on Microwave Theory and Techniques*, Vol. 44, No. 12, pp. 2680–2687, Dec. 1996.

[12] H.-C. Chang and K. A. Zaki, Evanescent-mode coupling of dual-mode rectangular waveguide filters, *IEEE Transactions on Microwave Theory and Techniques*, Vol. 39, No. 8, pp. 1307–1312, Aug. 1991.

[13] J.-F. Liang, X.-P. Liang, K. A. Zaki, and A. E. Atia, Dual-mode dielectric or air-filled rectangular waveguide filters, *IEEE Transactions on Microwave Theory and Techniques*, Vol. 42, No. 7, pp. 1330–1336, July 1994.

[14] M. Guglielmi, P. Jarry, E. Kerhervé, O. Roquebrun, and D. Schmitt, A new family of all-inductive dual-mode filters, *IEEE Transactions on Microwave Theory and Techniques*, pp. 1764–1769, Oct. 2001.

[15] M. Guglielmi, O. Roquebrun, P. Jarry, E. Kerhervé, M. Capurso, and M. Piloni, Low-cost dual-mode asymmetric filters in rectangular waveguide, *Proceedings of the IEEE International Microwave Symposium*, Vol. 3, pp. 1787–1790, May 2001.

[16] O. Roquebrun, P. Jarry, E. Kerhervé, and M. Guglielmi, Filtres micro-ondes en ligne possédant des zéros de transmission: circuit equivalent distribué, *11th National Microwave Days*, Arcachon, France, May 1999.

[17] A. E. Atia and A. E. Williams, New type of bandpass filters for satellites transponders, *COMSAT Technical Review*, Vol. 1, No. 1, pp. 21–43, 1971.

[18] W.-C. Tang and S. K. Chaudhuri, A true elliptic-function filter using triple-mode degenerate cavities, *IEEE Transactions on Microwave Theory and Techniques*, Vol. 32, No. 11, pp. 1449–1454, Nov. 1984.

[19] G. Lastoria, G. Gerini, M. Guglielmi, and F. Emma, CAD of triple-mode cavities in rectangular waveguide, *1998 Microwave and Guided Wave Letters*, Vol. 8, No. 10, pp. 339–341, Oct. 1998.

[20] R. R. Bonetti and A. E. Williams, Application of dual TM modes to triple- and quadruple-mode filters, *IEEE Transactions on Microwave Theory and Techniques*, Vol. 35, No. 12, pp. 1143–1149, Dec. 1987.

[21] L. Sheng-Li and L. Wei-Gan, A five mode single spherical cavity microwave filter, *MTT-S International Microwave Symposium Digest*, Vol. 92, No. 2, pp. 909–912, 1992.

[22] R. R. Bonetti and A. E. Williams, A hexa-mode bandpass filter, *1990 MTT-S International Microwave Symposium Digest 90.1*, Vol. 1, pp. 207–210.

[23] Y. C. M. Lim, R. F. Mostafavi, and D. Mirshekar-Syahkal, Small filters based on slotted cylindrical ring resonators, *IEEE Transactions on Microwave Theory and Techniques*, Vol. 49, No. 12, pp. 2369–2375, Dec. 2001.

[24] A. Enokihara, H. Nanba, T. Nakamura, T. Ishizaki, and T. Uwano, 26GHz TM_{110} mode dielectric resonator filter and duplexer with high-Q performance and compact configuration, *2002 MTT-S International Microwave Symposium Digest*, Vol. 3, pp. 1781–1784.

[25] S. B. Cohn, Microwave bandpass filters containing high-Q dielectric resonators, *IEEE Transactions on Microwave Theory and Techniques*, Vol. 6, No. 4, pp. 218–227, Apr. 1968.

[26] S. W. Chen, K. A. Zaki, and A. E. Atia, A single iris 8-pole dual mode dielectric resonator filter, *19th European Microwave Conference*, pp. 513–518, Sept. 1989.

[27] S. W. Chen and K. A. Zaki, A novel coupling method for dual-mode dielectric resonators and waveguide filters, *IEEE Transactions on Microwave Theory and Techniques*, Vol. 38, No. 12, pp. 1885–1893, Dec. 1990.

[28] T. Nishikawa, K. Wakino, K. Tsunoda, and Y. Ishikawa, Dielectric high-power bandpass filter using quarter-cut TE/sub 01delta/image resonator for cellular base stations, *IEEE Transactions on Microwave Theory and Techniques*, Vol. 35, No. 12, pp. 1150–1155, Dec. 1987.

[29] S. Moraud, S. Verdeyme, P. Guillon, Y. Latouche, S. Vigneron, and B. Theron, A new dielectric loaded cavity for high power microwave filtering, *1996 MTT-S International Microwave Symposium Digest*, Vol. 96, No. 2, pp. 615–618, 1996.

[30] Y. Tajima and Y. Sawayama, Design and analysis of a waveguide-sandwich microwave filter, *IEEE Transactions on Microwave Theory and Techniques*, Vol. 22, No. 9, pp. 839–841, Sept. 1974.

[31] Y. Konishi and K. Uenakada, The design of a bandpass filter with inductive strip-planar circuit mounted in waveguide, *IEEE Transactions on Microwave Theory and Techniques*, Vol. 22, No. 10, pp. 869–873, Oct. 1974.

[32] Y. C. Shih, T. Itoh, and L. Q. Bui, Computer-aided design of millimeter-wave E-plane filters, *1982 MTT-S International Microwave Symposium Digest*, Vol. 82, No. 1, pp. 471–473, 1982.

[33] J. Bornemann, Selectivity-improved E-plane filter for millimeter-wave applications, *Electronics Letters*, Vol. 27, No. 21, Oct. 1991.

[34] A. F. Sheta, K. Hettak, J. P. Coupez, C. Person, S. Toutain, and J. P. Blot, A new semi-lumped microwave filter structure, *1995 MTT-S International Microwave Symposium Digest*, pp. 383–386, 1995.

[35] S. B. Cohn, Design considerations for high-power microwave filters, *IEEE Transactions on Microwave Theory and Techniques*, Vol. 1, No. 7, pp. 154–162, Jan. 1959.

[36] S. B. Cohn, Parallel-coupled transmission line resonator filters, *IEEE Transactions on Microwave Theory and Techniques*, Vol. 6, No. 2, pp. 223–231, Apr. 1958.

[37] E. G. Cristal and S. Frankel, Hairpin-line and hybrid hairpin-line half-wave parallel-coupled-line filters, *IEEE Transactions on Microwave Theory and Techniques*, Vol. 20, No. 11, pp. 719–728, Nov. 1972.

[38] G. L. Matthaei, N. O. Fenzi, R. J. Forse, and S. M. Rohlfing, Hairpin-comb filters for HTS and other narrow-band applications, *IEEE Transactions on Microwave Theory and Techniques*, Vol. 45, No. 8, Aug. 1997.

[39] G. L. Matthaei, Interdigital band-pass filters, *IEEE Transactions on Microwave Theory and Techniques*, Vol. 10, No. 6, pp. 479–491, Nov. 1962.

[40] R. J. Wenzel, Exact theory of interdigital band-pass filters and related coupled band-pass structures, *IEEE Transactions on Microwave Theory and Techniques*, Vol. 13, No. 5, pp. 559–575, Sept. 1965.

[41] G. L. Matthaei, Comb-line band-pass filters of narrow or moderate bandwidth, *Microwave Journal*, Vol. 6, pp. 82–91, Aug. 1963.

[42] G. Torregrosa-Penalva, G. Lopez-Risueno, and J. I. Alonso, A simple method to design wideband electronically tunable combline filter, *IEEE Transactions on Microwave Theory and Techniques*, Vol. 50, No. 1, pp. 172–177, Jan. 2002.

[43] J. R. Lee, J. H. Cho, and S. W. Yun, New compact bandpass filter using microstrip/spl lambda//4 resonators with open stub inverter, *Microwave and Guided Wave Letters*, pp. 526–527, Dec. 2000.

[44] J. S. Hong and M. J. Lancaster, Couplings of microstrip square open-loop resonators for cross-coupled planar microwave filters, *IEEE Transactions on Microwave Theory and Techniques*, Vol. 44, No. 12, pp. 2099–2109, Dec. 1996.

[45] J. S. Hong and M. J. Lancaster, *Microstrip Filters for RF/Microwave Applications*, Wiley, Hoboken, NJ, 2001.

[46] J. S. Hong and M. J. Lancaster, Microstrip cross-coupled trisection bandpass filters with asymmetric frequency characteristics, *IEE Proceedings on Microwave Antenna Propagation*, Vol. 46, No. 1, Feb. 1999.

[47] J. A. Curtis and S. J. Fiedziuszko, Miniature dual mode microstrip filters, *MTT-International Microwave Symposium Digest 91.2*, Vol. 2, pp. 443–446, 1991.

[48] A. Cassinese, M. Barra, G. Panariello, and R. Vaglio, Multi-stage dual-mode cross-slotted superconducting filters for telecommunication application, *MTT-S International Microwave Symposium Digest*, Vol. 1, pp. 491–494, 2001.

[49] L. Roselli, L. Lucchini, and P. Mezzanotte, Novel compact narrow-band microstrip dual-mode resonator filters for 3G telecommunication systems, http://www.taconic-add.com/pdf/technicalarticles–bandpassfilter.pdf.

[50] E. O. Hammerstad, Equations for microstrip circuit design, *Proceedings of the European Microwave Conference*, Hamburg, Germany, pp. 268–272, 1975.

[51] W. Heinrich, A. Jentzsch, and G. Baumann, Millimeter-wave characteristics of flip-chip interconnects for multi-chip modules, *IEEE Transactions on Microwave Theory and Techniques*, Vol. 46, No. 2, pp. 2264–2268, Oct. 1998.

[52] http://www.vtt.fi/ele/research/ope/pdf/lahti_integrated_millimeter_wave_filters.pdf.

[53] S. Darfeuille, J. Lintignat, Z. Sassi, B. Barelaud, L. Billonnet, B. Jarry, H. Marie, P. Meunier, and P. Gamand, Novel design and analysis procedures for differential-based filters on silicon, *International Journal of RF and Microwave Computer-Aided Engineering*, Vol. 17, No. 1, pp. 49–55, Jan. 2007

[54] C. Lascaux, F. Rouchaud, V. Madrangeas, M. Aubourg, and P. Guillon, Planar Ka-band high temperature superconducting filters for space applications, *International Microwave Symposium Digest 01.1*, Vol. 1, pp. 487–490, 2001.

[55] G. L. Matthaei and G. L. Hey-Shipton, Concerning the use of high-temperature superconductivity in planar microwave filters, *IEEE Transactions on Microwave Theory and Techniques*, Vol. 42, No. 7, pp. 1287–1294, July 1994.

[56] R. E. Collin and F. J. Zucker, *Antenna Theory*, McGraw-Hill New York, 1969.

[57] F. R. Yang, R. Coccioli, Y. Qian, and T. Itoh, Analysis and application of coupled microstrips on periodically patterned ground plane, *2000 MTT-S International Microwave Symposium Digest*, Vol. 3, pp. 1529–1532, 2000.

[58] P. V. Wright, A review of SAW resonator filter technology, *IEEE Ultrasonics Symposium*, pp. 29–39, 1992.

[59] K. M. Lakin, Bulk acoustic wave coupled resonator filters, *Proceedings of the IEEE Frequency Control Symposium and PDA Exhibition*, pp. 8–14, May 2002.

[60] A. Shirakawa, J. M. Pham, P. Jarry, and E. Kerhervé, Bulk acoustic wave coupled resonator filters synthesis methodology, *35th European Microwave Conference (EuMC 2005)*, Paris, pp. 459–462, Oct. 3–7, 2005.

[61] N. I. Dib, W. P. Harokopus, P. B. Katehi, C. C. Ling, and G. M. Rebeiz, Study of a novel planar transmission line, *MTT-S International Microwave Symposium Digest*, Vol. 2, pp. 623–626, 1991.

[62] C. Y. Chi and G. M. Rebeiz, A low-loss 20GHz micromachined bandpass filter, *International Microwave Symposium Digest 95.3*, Vol. 3, pp. 1531–1534, 1995.

[63] T. J. Ellis and G. M. Rebeiz, MM-wave tapered slot antennas on micromachined photonic bandgap dielectrics, *MTT-S International Microwave Symposium Digest*, Vol. 2, pp. 1157–1160, 1996.

[64] H. E. Brenner, Use a computer to design suspended substrate ICs, *Microwaves Journal*, pp. 38–45, Sept. 1968.

[65] E. Yamashita and K. Atsuki, Strip line with rectangular outer conductor and three dielectric layers, *IEEE Transactions on Microwave Theory and Techniques*, Vol. 18, No. 5, pp. 238–244, May 1970.

[66] W. Schwab, F. Boegelsack, and W. Menzel, Multilayer suspended stripline and coplanar line filters, *Transactions on Microwave Theory and Techniques*, pp. 1403–1407, July 1994.

[67] C. Collado, J. Pozo, J. Mateu, and J. M. O'Callaghan, Compact duplexer with dual loop resonators, *35th European Microwave Conference*, Paris, pp. 109–112, Oct. 3–7, 2005.

[68] C. Y. Tsai and J. T. Kuo, A new miniaturized dual-mode loop filter using coupled compact miniaturized hairpin resonators, *Proceedings of the Asia Pacific Microwave Conference (APMC 2004)*, New Delhi, India, Dec. 15–18, 2004.

[69] K. K. Sun and K. W. Tam, A novel compact dual-mode bandpass filter with meander open-loop arms, *2004 MTT-S International Microwave Symposium Digest*, Fort Worth, TX, pp. 1479–1482, 2004.

CHAPTER 2

IN-LINE SYNTHESIS OF PSEUDO-ELLIPTIC FILTERS

2.1 Introduction 35
2.2 Approximation and synthesis 36
2.3 Chebyshev filters 36
2.4 Pseudo-elliptic filters 37
2.4.1 Pseudo-elliptic characteristic function 38
2.4.2 Pseudo-elliptic transfer functions 40
2.4.3 Cross-coupled and in-line prototypes for asymmetrical responses 41
2.4.4 Analysis of the in-line prototype elements 45
2.4.5 Synthesis algorithm for pseudo-elliptic lowpass in-line filters 49
2.4.6 Frequency transformation 50
2.5 Prototype synthesis examples 52
2.6 Theoretical coupling coefficients and external quality factors 59
References 61

2.1 INTRODUCTION

The design of the fractal filters described in this book starts by defining a suitable synthesis procedure to compute the theoretical elements of an electrical prototype network that can provide specified Chebysev and pseudo-elliptic responses. Using

Design and Realizations of Miniaturized Fractal RF and Microwave Filters,
By Pierre Jarry and Jacques Beneat

the electrical prototype network, the theoretical coupling coefficients between resonators and theoretical external quality factors can then be computed. The theoretical coupling coefficients and external quality factor are then used to define the final physical parameters of the fractal filters. In this chapter, the synthesis procedure of the electrical prototype network for pseudo-elliptic responses, and formulas for computing the theoretical coupling coefficients and external quality factors, are provided.

2.2 APPROXIMATION AND SYNTHESIS

There are several approximations to the ideal lowpass filter. Butterworth, Chebyshev, elliptic, and pseudo-elliptic functions approximate the ideal lowpass magnitude response, and the Bessel and equidistant linear-phase functions approximate the ideal lowpass phase response. The bandpass magnitude response can replicate the lowpass magnitude response by applying a frequency transformation on the approximation functions [1]. The ideal lowpass and bandpass magnitude responses are shown in Figure 2.1.

The synthesis of a filter consists in defining a realizable transfer function that exhibits the same characteristics as the approximation functions. An electric prototype circuit can then be defined from the transfer function using various techniques, such as successive extractions of circuit elements. The electrical prototype circuit can then be used as a starting point for designing microwave filters. The design will involve defining the physical parameters of the microwave structure so that it can produce the same filtering characteristics as the prototype circuit.

2.3 CHEBYSHEV FILTERS

The Chebyshev transmission function is given by

$$|S_{21}(j\omega)|^2 = \frac{1}{1 + \varepsilon^2 T_n^2(\omega)}$$

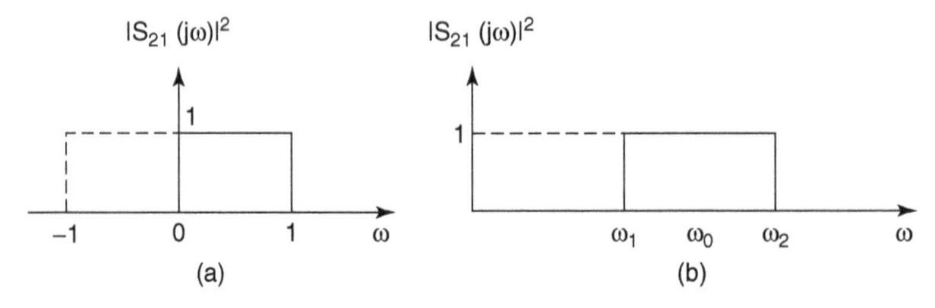

Figure 2.1 Ideal lowpass and bandpass magnitude responses.

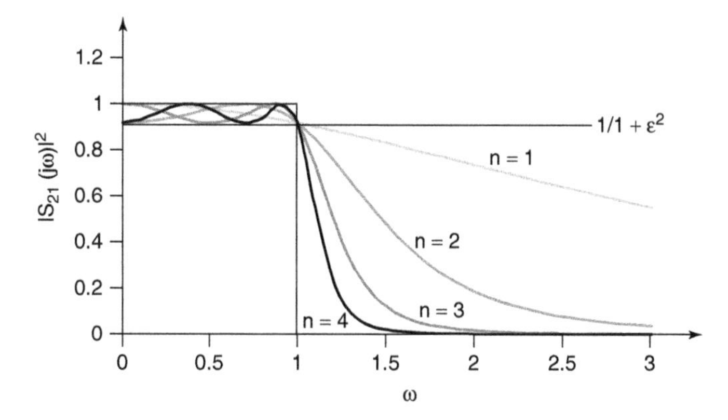

Figure 2.2 Transmission characteristic of Chebyshev filters.

where ε characterizes the ripple in the band, n is the filter's order, and $T_n(\omega)$ is a function defined as follows:

$$T_n(\omega) = \begin{cases} \cos(n \text{ arccos } \omega) & \omega \leq 1 \\ \text{ch}(n \text{ argch } \omega) & \omega > 1 \end{cases}$$

Figure 2.2 shows the transmission function when the order n varies from 1 to 4.

The lowpass prototype circuit is composed of a ladder made of inductors and capacitors terminated on 1-Ω terminations. A more suitable prototype circuit for microwave applications is a ladder made of either inductors or capacitors using impedance or admittance inverters [3]. The bandpass characteristic is obtained by applying a frequency transformation on the lowpass transfer function. This operation maps the lowpass passband edges -1 and $+1$ to the bandpass passband edges ω_1 and ω_2. The transformation is given below, where s denotes the complex Laplace variable and ω_0 represents the center frequency of the bandpass filter.

$$s \rightarrow \alpha \left(\frac{s}{\omega_0} + \frac{\omega_0}{s} \right)$$

$$\omega_0 = \sqrt{\omega_1 \omega_2} \qquad \alpha = \frac{\omega_0}{\omega_2 - \omega_1}$$

In the prototype circuit, an inductor is transformed into a series inductor–capacitor combination, and a capacitor is transformed into a parallel inductor–capacitor. Terminations and impedance or admittance inverters remain unchanged.

2.4 PSEUDO-ELLIPTIC FILTERS

For applications such as multiplexing, the selectivity of the filter must be high around the passband of the filter. To improve the selectivity of a Chebyshev filter, one must increase the order of the filter. This increases the complexity and size

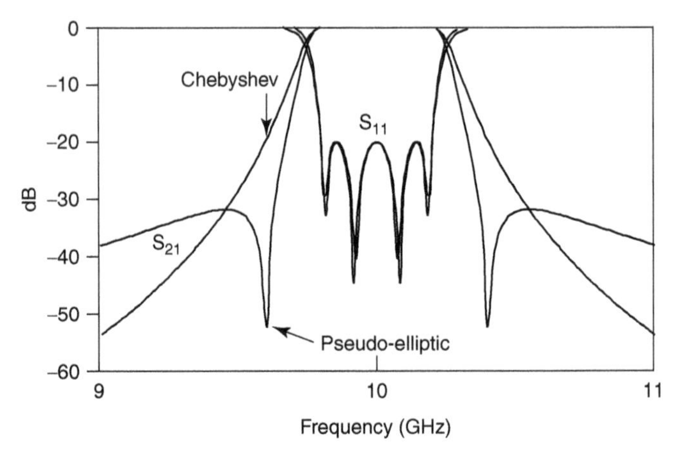

Figure 2.3 Comparison of Chebyshev and pseudo-elliptic responses.

of the filter and can increase the insertion losses. A preferred technique is to insert a few transmission zeros near the passband.

A well-known function that presents transmission zeros is the elliptic function. Unfortunately, this function has a large number of transmission zeros, far more than what is actually needed for microwave filters. A function that can maintain the equiripple behavior of the Chebyshev function in the passband while improving the selectivity with only a few transmission zeros is the pseudo-elliptic function, also referred to as the generalized Chebyshev function.

Figure 2.3 shows how the pseudo-elliptic function provides improved selectivity over the Chebyshev function. The two filters are designed for a center frequency of 10 GHz and a bandwidth of 400 MHz. The first filter uses a fourth-order Chebyshev function with −20 dB of return loss, and the second filter uses a fourth-order pseudo-elliptic function with −20 dB return loss and two transmission zeros. Clearly, the selectivity near the passband is greatly improved by the transmission zeros.

2.4.1 Pseudo-elliptic Characteristic Function

The pseudo-elliptic transmission function is based on the Chebyshev function [3–6] and is expressed as

$$|S_{21}(j\omega)|^2 = \frac{1}{1 + \varepsilon^2 F_n^2(j\omega)}$$

where the function $F_n(j\omega)$ is defined as

$$F_n(\omega) = \begin{cases} \cos\left[(n - n_z)\cos^{-1}\omega + \sum_{i=1}^{n_z} \cos^{-1}\dfrac{1 - \omega\omega_i}{\omega - \omega_i}\right] & \text{inside the passband} \\ \cosh\left[(n - n_z)\cosh^{-1}\omega + \sum_{i=1}^{n_z} \cosh^{-1}\dfrac{1 - \omega\omega_i}{\omega - \omega_i}\right] & \text{outside the passband} \end{cases}$$

n is the order of the filter and represents the number of Chebyshev ripples in the passband, n_z is the number of transmission zeros, and the ω_i are the radian frequencies of the transmission zeros.

The function $F_n(\omega)$ is defined as the ratio of two polynomials in ω:

$$F_n(\omega) = \frac{E_n^+ + E_n^-}{2Q_n}$$

with

$$Q_n(\omega) = \prod_{i=1}^{n} \left(1 - \frac{\omega}{\omega_i}\right)$$

$$E_n^+(\omega) = \prod_{i=1}^{n} \left[\left(\omega - \frac{1}{\omega_i}\right) + \omega' \sqrt{1 - \frac{1}{\omega_i^2}}\right]$$

$$E_n^-(\omega) = \prod_{i=1}^{n} \left[\left(\omega - \frac{1}{\omega_i}\right) - \omega' \sqrt{1 - \frac{1}{\omega_i^2}}\right]$$

$$\omega' = \sqrt{\omega^2 - 1}$$

The coefficients of $F_n(\omega)$ are defined by calculating the functions $E_n^+(\omega)$ and $E_n^-(\omega)$. These polynomial functions are calculated iteratively. The initial parameters that generate these polynomial functions are the passband ripple, the filter order, and the radian frequencies of the transmission zeros. If we write

$$E_n^+(\omega) = P_n(\omega) + M_n(\omega)$$

$$E_n^-(\omega) = P_n(\omega) - M_n(\omega)$$

then

$$F_n(\omega) = \frac{P_n}{Q_n}$$

The polynomials P_n and M_n are generated as follows:

$$P_1(\omega) = -\frac{1}{\omega_1} + \omega$$

$$M_1(\omega) = \omega' \sqrt{1 - \frac{1}{\omega_1^2}}$$

$$P_2(\omega) = \omega P_1(\omega) - \frac{P_1(\omega)}{\omega_2} + \omega' \sqrt{1 - \frac{1}{\omega_2^2}}\, M_1(\omega)$$

$$M_2(\omega) = \omega' M_1(\omega) - \frac{M_1(\omega)}{\omega_2} + \omega' \sqrt{1 - \frac{1}{\omega_2^2}} P_1(\omega)$$

$$P_n(\omega) = \omega P_{n-1}(\omega) - \frac{P_{n-1}(\omega)}{\omega_n} + \omega' \sqrt{1 - \frac{1}{\omega_n^2}} M_{n-1}(\omega)$$

$$M_n(\omega) = \omega' M_{n-1}(\omega) - \frac{M_{n-1}(\omega)}{\omega_n} + \omega' \sqrt{1 - \frac{1}{\omega_n^2}} P_{n-1}(\omega)$$

The polynomial Q_n is generated as follows:

$$Q_1(\omega) = 1 - \frac{\omega}{\omega_1}$$

$$Q_2(\omega) = \left(1 - \frac{\omega}{\omega_2}\right) Q_1(\omega)$$

The procedure is repeated until the last transmission zero is accounted for:

$$Q_{n_z}(\omega) = \left(1 - \frac{\omega}{\omega_{n_z}}\right) Q_{n_z-1}(\omega)$$

The transmission and reflection functions can be expressed in terms of the polynomials P_n and Q_n as follows:

$$|S_{21}(j\omega)|^2 = \frac{1}{1 + \varepsilon^2 F_n^2(j\omega)} = \frac{Q_n^2(j\omega)}{Q_n^2(j\omega) + \varepsilon^2 P_n^2(j\omega)}$$

$$|S_{11}(j\omega)|^2 = 1 - |S_{21}(j\omega)|^2 = \frac{\varepsilon^2 P_n^2(j\omega)}{Q_n^2(j\omega) + \varepsilon^2 P_n^2(j\omega)}$$

The transmission transfer function can then be written as

$$S_{21}(s) = \frac{P_n(s)}{\varepsilon R_n(s)}$$

2.4.2 Pseudo-elliptic Transfer Functions

After determining the polynomials $P_n(j\omega)$ and $Q_n(j\omega)$, it is possible to define the transmission transfer functions $S_{21}(s)$ and $S_{11}(s)$. One notes that both transfer functions will have the same denominator polynomial. This polynomial is defined by first computing the roots in ω of $Q_n^2(j\omega) + \varepsilon^2 P_n^2(j\omega)$. This can be done using the function roots in MATLAB, for instance. Then the roots in s are obtained by using the relation $s \to j\omega$. Finally, only the roots in the left half-plane (real part negative) are retained for stability issues. For the numerator polynomials, one can simply replace ω according to $s \to j\omega$ in the polynomials$P_n(j\omega)$ and

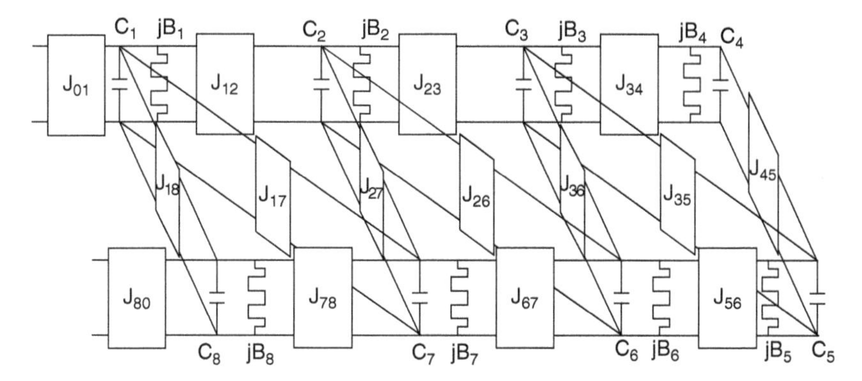

Figure 2.4 Cross-coupled lowpass prototype network.

$Q_n(j\omega)$. One should note that when the zeros chosen provide an asymmetrical response around zero, the transfer function can have complex coefficients. This can be the case, for example, when one is only interested in a single transmission zero, on either the left or right of the passband.

2.4.3 Cross-Coupled and In-Line Prototypes for Asymmetrical Responses

A generalized lowpass prototype that can generate asymmetrical pseudo-elliptic filtering characteristics, first presented by Mohamed and Lind [7], is shown in Figure 2.4. This structure, often referred to as a *cross-coupled network*, is composed of frequency-independent admittance inverters J_{ij}, capacitances C_i, and susceptances B_i. The susceptances are imaginary elements since they do not exist in reality. They are, however, necessary in the case of asymmetrical responses to account for the complex-valued coefficients of the transfer function. The admittance inverters represent the couplings between the resonators represented by the shunt C_i and B_i combinations. The elements of the cross-coupled network can be defined from the transfer function using a technique based on the associated *ABCD* or chain matrix of the transfer function [4]. This technique is rather tedious, however, and requires going back and forth between the start and the end of the structure while extracting the various elements.

An alternative lowpass prototype, the in-line network [3,9,10], is shown in Figure 2.5. It is composed of parallel resonators with admittance inverters [Figure 2.5(a)] or of series resonators with impedance inverters [Figure 2.5(b)]. The admittance or impedance inverters create the couplings between nonadjacent resonators to produce the transmission zeros. The two in-line prototypes are equivalent when taking $L_i = C_i$, $X_i = B_i$, and $K_i = -1/J_i$. Figure 2.6 provides the in-line bandpass prototypes associated with the in-line lowpass prototypes of Figure 2.5. The inverters remain the same since they are frequency independent, while a C_i–B_i resonator is replaced by a parallel capacitor–inductor resonator, and an L_i–X_i resonator is replaced by a series capacitor–inductor

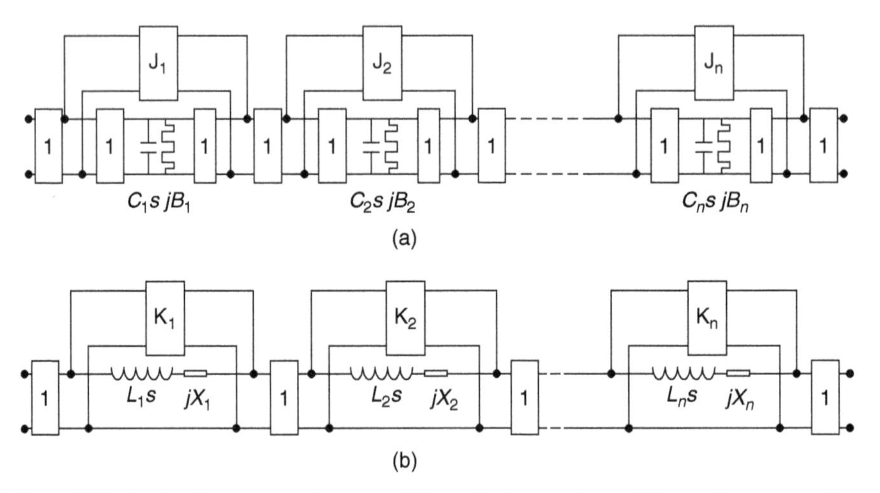

Figure 2.5 In-line lowpass prototypes: (a) parallel resonators with admittance inverters; (b) series resonators with impedance inverters.

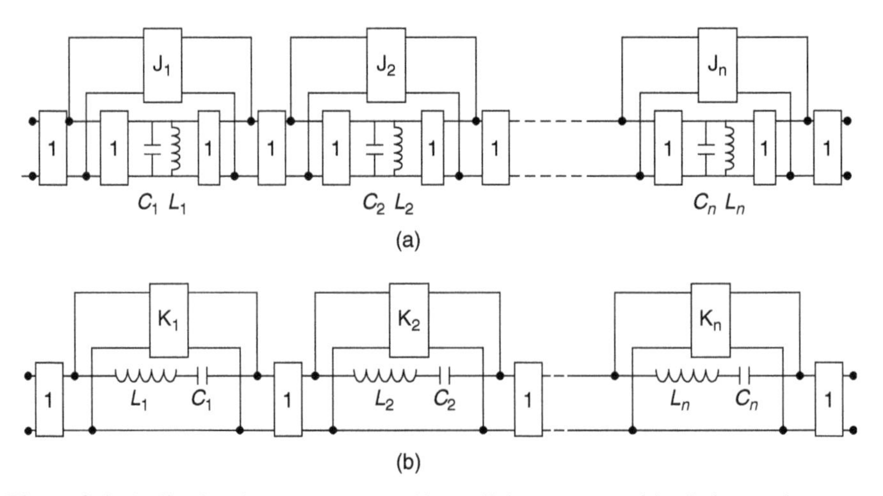

Figure 2.6 In-line bandpass prototypes: (a) parallel resonators with admittance inverters; (b) series resonators with impedance inverters.

resonator. Contrary to the lowpass network, the bandpass network is composed of realizable elements.

Since every resonator in association with an inverter (cross coupling) will create a transmission zero, it is possible to create an nth-order filter with n transmission zeros. We show later how the fundamental element of the in-line prototype shown in Figure 2.7 can produce a transmission zero [8,9]. With reference to the symmetry plane shown in Figure 2.7(b), it is possible to apply the Bartlett bisection theorem and define the even and odd modes of the

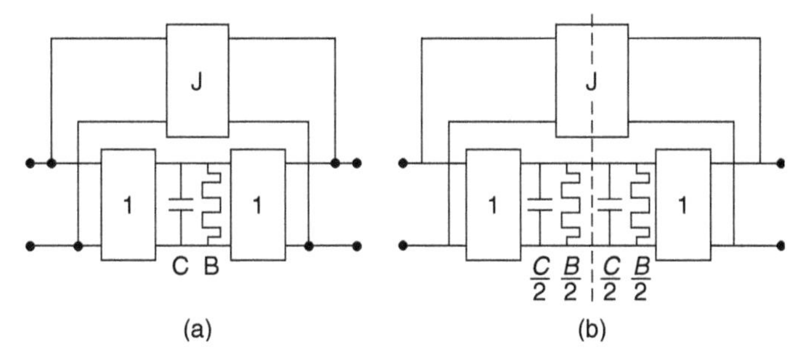

Figure 2.7 (a) Fundamental element of an in-line lowpass prototype; (b) symmetry plane used for the demonstration.

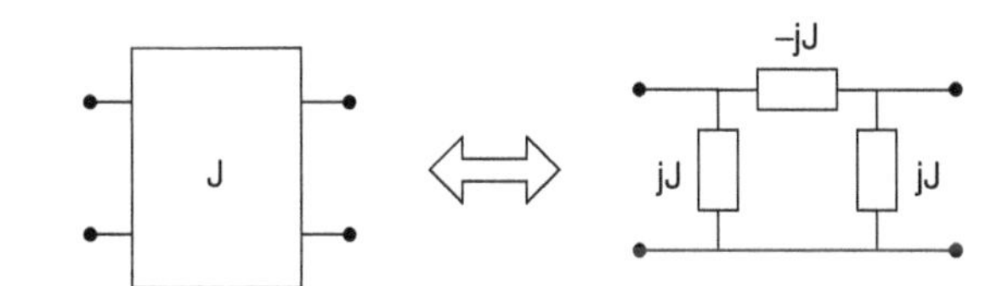

Figure 2.8 Equivalent π circuit of an admittance inverter.

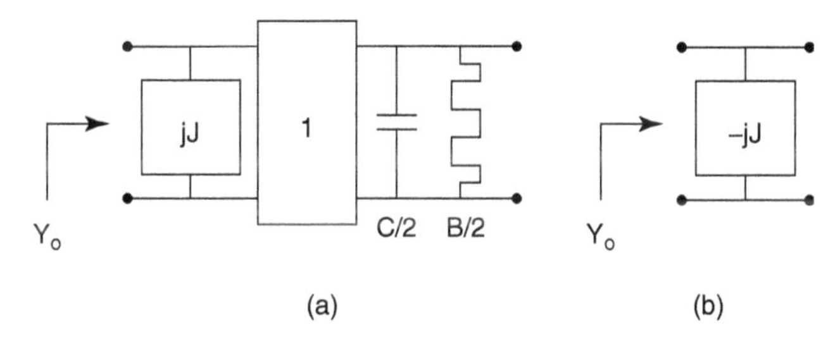

Figure 2.9 Determining the (a) even and (b) odd admittances.

circuit [8–10]. In addition, the admittance inverter is replaced by its equivalent π circuit, where jJ represents the inverter's admittance, as shown in Figure 2.8.

The even-mode admittance is determined by placing an open circuit in the symmetry plane of the elementary element. From Figure 2.9(a) we have

$$Y_e = jJ + \frac{1}{(C/2)p + j(B/2)}$$

The odd-mode admittance is obtained by placing a short circuit in the symmetry plane of the fundamental element. From Figure 2.9(b) we have

$$Y_o = -jJ$$

The transfer function of the fundamental element is then given by

$$S_{21}(s) = \frac{Y_e(s) - Y_o(s)}{[1 + Y_e(s)][1 + Y_o(s)]}$$

Replacing the even and odd admittances in the formula above, we have

$$S_{21}(s) = \frac{jJ(Cs + jB) + 1}{(1 - jJ)\{1 + (1 + jJ)[(C/2)s + j(B/2)]\}}$$

The transmission function $S_{21}(s)$ has a zero at the finite frequency ω_i given by

$$\omega_i = -\frac{1}{C}\left(B - \frac{1}{J}\right)$$

One can change the transmission zero frequency to $-\omega_i$ by replacing B by $-B$ and J by $-J$. In general, the transmission zero frequency will be given by

$$\omega_i = \pm\frac{1}{C}\left|B - \frac{1}{J}\right|$$

This property provides us with two possible locations for the transmission zero. It can be located either on the right or left of the passband, as shown in Figure 2.10.

The in-line network is capable of realizing transfer functions with asymmetrical frequency responses where the transmission zeros can be placed either on the left or right of the passband according to the sign of the coupling created by the admittance inverter. The next step consists of determining the values of capacitances C_i, susceptances B_i, and admittance inverters J_{ij} from the transfer functions $S_{21}(s)$ and $S_{11}(s)$. But first, further analysis of the elements of the in-line prototype is needed.

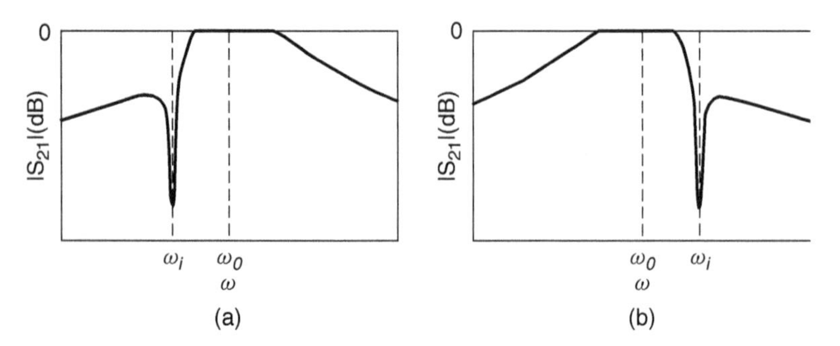

Figure 2.10 Possible transmission zero locations.

2.4.4 Analysis of the In-Line Prototype Elements

The fundamental element of the in-line network is not easy to extract under its current form. Using the equivalent π circuit of the admittance inverter of Figure 2.8, the fundamental element of the in-line prototype can be redrawn as the circuit in Figure 2.11 [3,11–13]. By replacing the parallel admittance by an equivalent series impedance [Figure 2.12(a)], the fundamental element can be drawn as shown in Figure 2.12(b), where $L = C$ and $X = B$. It is important to notice that the series elements Ls and jX represent impedances, whereas the coupling $-jJ$ represents an admittance. Figure 2.13 shows the final modifications

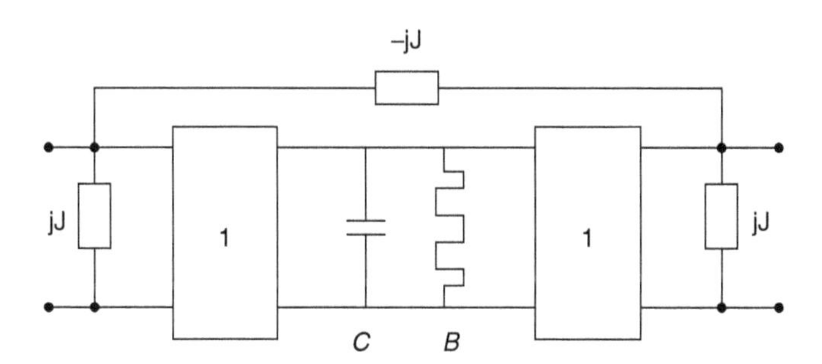

Figure 2.11 First equivalent circuit of the fundamental element.

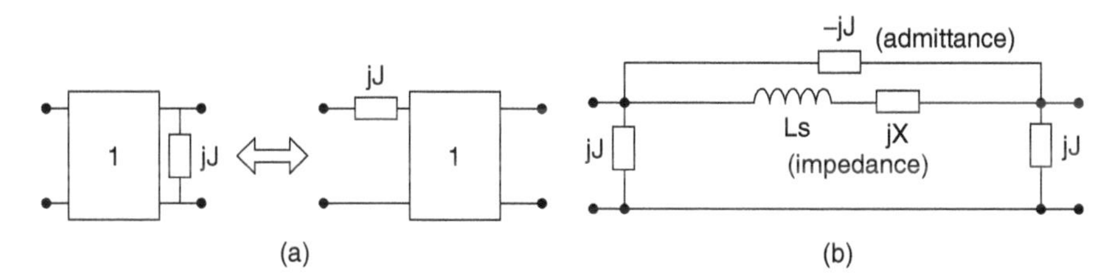

Figure 2.12 (a) Parallel admittance/series impedance equivalence; (b) second equivalent circuit of the fundamental element.

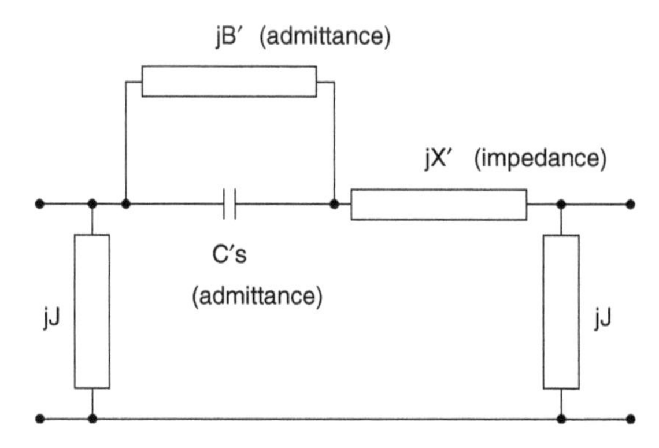

Figure 2.13 Equivalent circuit of the fundamental element.

to the representation of the fundamental element. The circuits of Figures 2.12(b) and 2.13 would be the same if the impedance $Z(s)$ seen between the two shunt jJ's were the same. This is true if

$$J = \frac{1}{X'}$$
$$L = X'^2 C'$$
$$X = (X'B' - 1)X'$$

The impedance of the equivalent circuit is given by

$$Z(s) = \frac{1}{C's + jB'} + jX'$$

This impedance should become infinite at the frequency ω_i of the transmission zero. Therefore, we must have

$$C's + jB'|_{s=j\omega_i} = jC'\omega_i + jB' = 0$$

This provides the additional relation

$$B' = -\omega_i C'$$

Using the relations above, the equivalent circuit can be expressed mainly in terms of J, C', and ω_i, as shown in Figure 2.14. The element values of the fundamental element can then be computed from the element values of the equivalent circuit using

$$J = \frac{1}{X'}$$
$$C = X'^2 C'$$
$$B = (X'B' - 1)X'$$

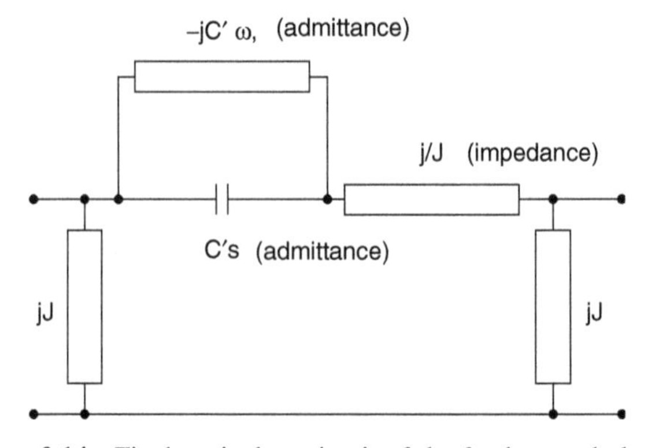

Figure 2.14 Final equivalent circuit of the fundamental element.

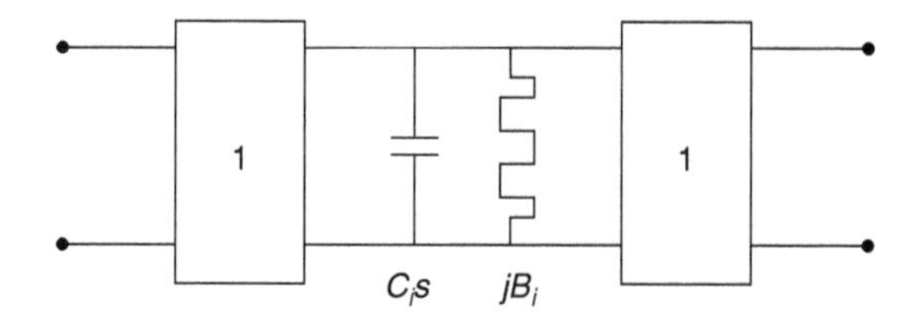

Figure 2.15 Simple resonator.

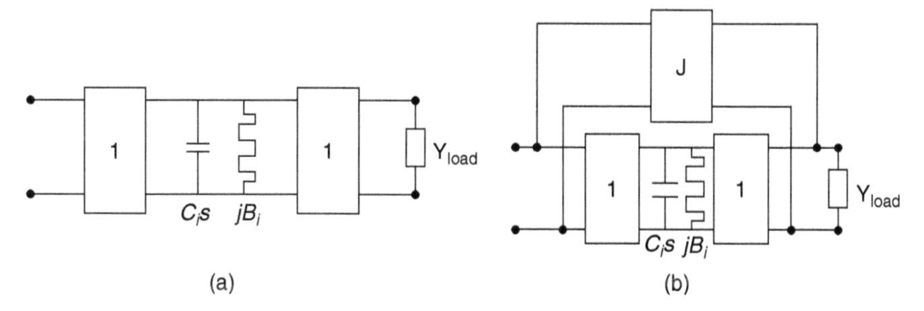

Figure 2.16 Elements to synthesize: (a) simple resonator; (b) fundamental element.

The extraction of the prototype elements is the first step in the generalized synthesis algorithm. There are two main elements that can form the lowpass in-line prototype: the fundamental element of Figure 2.7(a) and the simple resonator of Figure 2.15. This simple resonator is used to introduce additional resonance in the passband without adding a new transmission zero. It is simply the fundamental element where the coupling inverter has been removed.

The synthesis of the in-line network requires that we be able to synthesize a $C-B$ resonator that generates a pole of the transfer function [Figure 2.16(a)], and to synthesize the fundamental element that generates a pole and a transmission zero [Figure 2.16(b)].

Synthesis Steps for a Simple Resonator The admittance and impedance of a simple resonator are given by

$$Y(s) = \frac{1}{C_i s + jB_i + 1/Y_{\text{load}}}$$

$$Z(s) = C_i s + jB_i + 1/Y_{\text{load}}$$

Thus, to extract a capacitor, we can divide $Z(s)$ by s and take the limit when s approaches infinity:

$$C_i = \left.\frac{Z(s)}{s}\right|_{s\to\infty}$$

The quantity $C_i s$ is then subtracted from the impedance:

$$Z(s) - C_i s = jB_i + \frac{1}{Y_{\text{load}}}$$

The susceptance B_i and the load Y_{load} can then be found from the imaginary and real parts of the remainder, respectively.

Synthesis Steps for a Fundamental Element The admittance of the equivalent circuit of the fundamental element is given by

$$Y(s) = jJ + \cfrac{1}{\cfrac{1}{C_i'(s - j\omega_i)} + \cfrac{j}{J} + \cfrac{1}{jJ + Y_{\text{load}}}}$$

By letting $s \to j\omega_i$, the coupling J can be extracted:

$$jJ = Y(s)|_{s=j\omega_i}$$

The effects of the inverter are then removed to obtain a residue impedance as follows:

$$Z(s) = \frac{1}{Y(s) - jJ} - \frac{j}{J} = \frac{1}{C_i'(s - j\omega_i)} + \frac{1}{jJ + Y_{\text{load}}}$$

By multiplying both sides by $s - j\omega_i$, we obtain

$$(s - j\omega_i)Z(s) = \frac{1}{C_i'} + \frac{s - j\omega_i}{jJ + Y_{\text{load}}}$$

The capacitance C_i' can then be extracted by letting $s \to j\omega_i$:

$$\frac{1}{C_i'} = (s - j\omega_i)Z(s)|_{s=j\omega_i}$$

The effects of the capacitance are then removed to form a new residue impedance as follows:

$$Z_1(s) = \frac{1}{Y(s) - jJ} - \frac{j}{J} - \frac{1}{C_i'(s - j\omega_i)} = \frac{1}{jJ + Y_{\text{load}}}$$

The load Y_{load} can then be obtained from the real part of the remainder. The elements of the fundamental element are then computed from the extracted elements of the equivalent circuit using

$$C_i = \frac{C_i'}{J^2}$$

$$B_i = -\frac{J + C_i'\omega_i}{J^2}$$

2.4.5 Synthesis Algorithm for Pseudo-elliptic Lowpass In-line Filters

To illustrate the synthesis procedure, the example of a third-order pseudo-elliptic filter with one transmission zero is used [3]. For this case, two simple resonators provide two poles, and one fundamental element provides one pole and one transmission zero. The prototype circuit, shown in Figure 2.17, provides three poles and one transmission zero.

First each fundamental element is replaced by its equivalent circuit of Figure 2.14. This provides the intermediate circuit of Figure 2.18. The input admittance of the intermediate circuit of Figure 2.18 is then defined.

$$Y(s) = C_1 s + jB_1 + jJ_2 + \cfrac{1}{\cfrac{j}{J_2} + \cfrac{1}{C_2'(s - j\omega_2)} + \cfrac{1}{C_3 s + jB_3 + jJ_2 + 1/Y_{\text{load}}}}$$

We begin by extracting the capacitance and susceptance of the first simple resonator:

$$C_1 = \left.\frac{Y(s)}{s}\right|_{s\to\infty}$$

$$Y_1(s) = Y(s) - C_1 s$$

$$jB_1 = Y_1(s)|_{s\to\infty} = jB_1 + jJ_2 - jJ_2$$

We then extract the fundamental element:

$$Y_2(s) = Y_1(s) - jB_1$$

$$jJ_2 = Y_2(s)|_{s=j\omega_2}$$

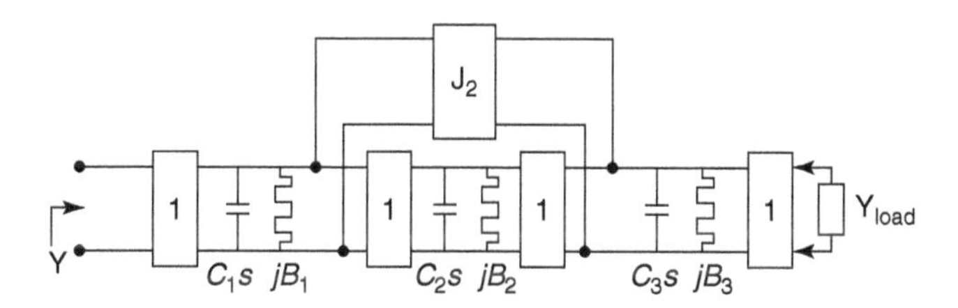

Figure 2.17 In-line prototype for a third-order pseudo-elliptic filter with one transmission zero.

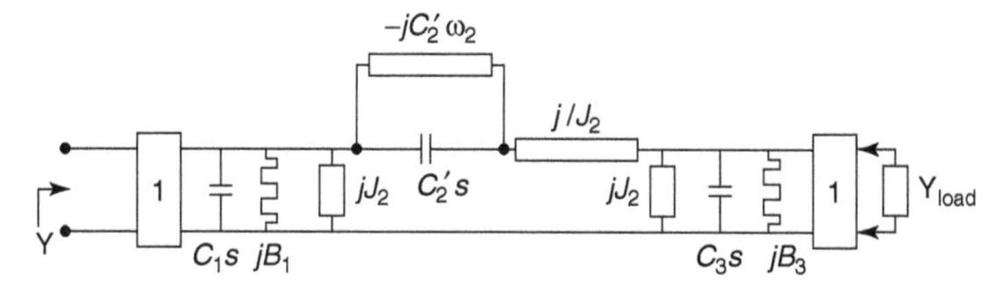

Figure 2.18 Third-order lowpass prototype with an intermediate circuit.

$$Z_2(s) = \frac{1}{Y_2(s) - jJ_2} - \frac{j}{J_2} = \frac{1}{C_2'(s - j\omega_2)} + \frac{1}{C_3 s + jB_3 + jJ_2 + 1/Y_{\text{load}}}$$

$$(s - j\omega_2) Z_2(s) = \frac{1}{C_2'} + \frac{s - j\omega_2}{C_3 s + jB_3 + jJ_2 + 1/Y_{\text{load}}}$$

$$\frac{1}{C_2'} = (s - j\omega_2) Z_2(s)|_{s=j\omega_2}$$

Then we extract the final simple resonator:

$$Z_3(s) = Z_2(s) - \frac{1}{C_2'(s - j\omega_2)} = \frac{1}{C_3 s + jB_3 + jJ_2 + 1/Y_{\text{load}}}$$

$$Y_3(s) = \frac{1}{Z_3(s)} = C_3 s + j(B_3 + J_2) + 1/Y_{\text{load}}$$

$$C_3 = \left.\frac{Y_3(s)}{s}\right|_{s\to\infty}$$

$$Y_4(s) = Y_3(s) - C_3 s - jJ_2 = jB_3 + \frac{1}{Y_{\text{load}}}$$

The susceptance B_3 and the load Y_{load} can then be found from the imaginary and real parts of the remainder, respectively. The synthesis method can be generalized to higher-order cases using the flowchart in Figure 2.19.

2.4.6 Frequency Transformation

The traditional lowpass-to-bandpass transformation is as follows:

$$s \to \alpha\left(\frac{s}{\omega_0} + \frac{\omega_0}{s}\right) \qquad \omega_0 = \sqrt{\omega_1 \omega_2} \qquad \alpha = \frac{\omega_0}{\omega_2 - \omega_1}$$

where $1/\alpha$ is the relative or fractional bandwidth and ω_0 is the central frequency of the filter. This transformation is applied to the lowpass transfer function. It has the effect of mapping the lowpass passband edges -1 and $+1$ to the bandpass passband edges ω_1 and ω_2 and preserves the magnitude response characteristics. This transform changes a capacitor to a parallel capacitor–inductor resonator and an inductor to a series capacitor–inductor resonator. This transform does not change the admittance or impedance inverters since they are independent of frequency. This is how bandpass prototype elements are defined traditionally from lowpass prototype elements. However, in the case of cross-coupled or in-line prototypes, one must examine more carefully what happens due to the presence of the imaginary susceptance B_i.

We decide that this transformation will transform the capacitor C_i in parallel with the susceptance B_i into a capacitor C_{ai} in parallel with an inductor L_{ai} as shown in Figure 2.20 [2,3]. We first match the transformed admittance with the admittance of the desired circuit of Figure 2.20.

$$\alpha\left(\frac{\omega}{\omega_0}-\frac{\omega_0}{\omega}\right)C_i+B_i=C_{ai}\omega-\frac{1}{L_{ai}\omega}$$

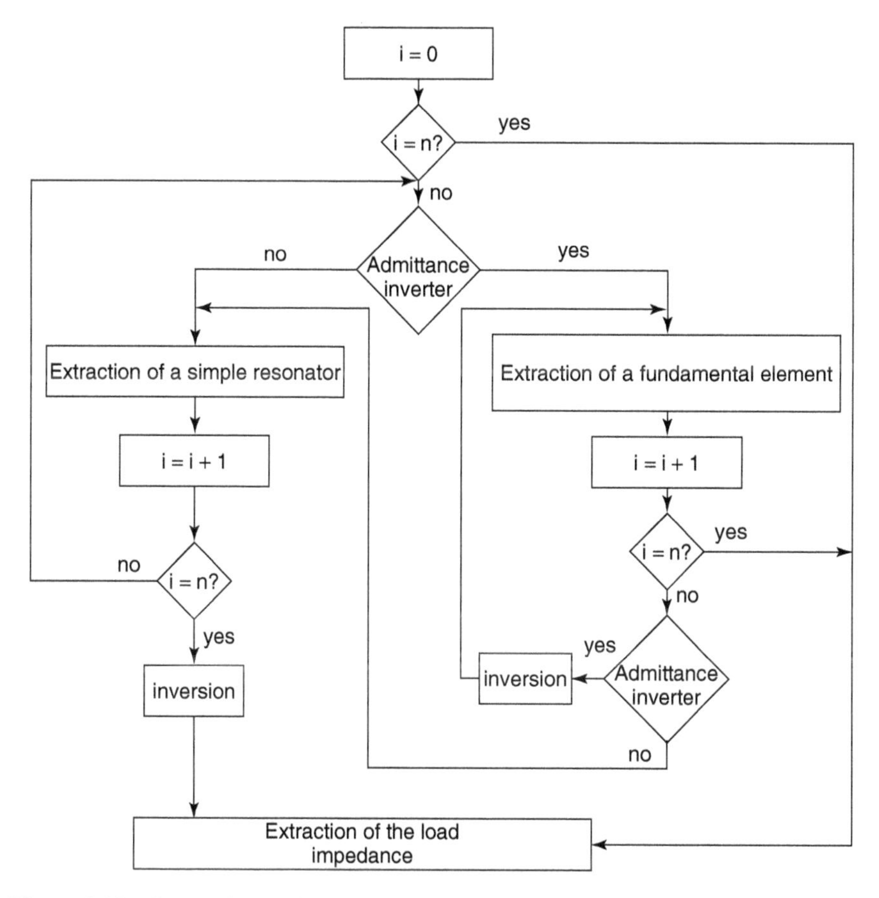

Figure 2.19 Synthesis algorithm for pseudo-elliptic lowpass in-line prototype networks.

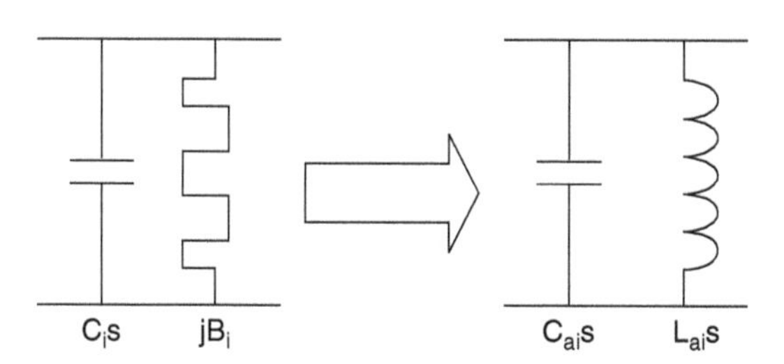

Figure 2.20 Lowpass-to-bandpass transformation.

Another equation is needed, however. We then also match the derivative with regard to ω of the transformed admittance with the derivative of the admittance of the circuit desired.

$$\alpha\left(\frac{1}{\omega_0}+\frac{\omega_0}{\omega^2}\right)C_i = C_{ai}+\frac{1}{L_{ai}\omega^2}$$

The two equations must be true for every ω, and in particular for $\omega = \omega_0$, the center frequency of the filter. In this case the bandpass elements are defined from the lowpass elements using

$$C_{ai} = \frac{1}{2}\left(\frac{2\alpha C_i + B_i}{\omega_0}\right)$$

$$L_{ai} = \frac{2}{\omega_0(2\alpha C_i - B_i)}$$

Note that this is a narrowband technique. As one gets farther from the center frequency, the bandpass circuit will not produce exactly the same response as the bandpass transfer function.

2.5 PROTOTYPE SYNTHESIS EXAMPLES

To illustrate the synthesis of asymmetrical pseudo-elliptic filters, several examples are provided. The first example [14,15] consists of a third-order filter with a transmission zero on the right side of the passband with the following specifications:

- *Central frequency:* 11.58 GHz
- *Bandwidth:* 80 MHz
- *Transmission zero frequency:* 11.72 GHz
- *Return loss:* 20 dB

The lowpass prototype circuit for this filter is shown in Figure 2.21. The first step is to define the location of the transmission zero for the lowpass prototype. It is given by

$$\omega_Z = \alpha\left(\frac{\omega_{pz}}{\omega_0}-\frac{\omega_0}{\omega_{pz}}\right)$$

where ω_{pz} is the location of the bandpass zero (e.g., $2\pi \times 11.72 \times 10^6$ rad/s). In this case we find that $\omega_z = 3.479$ rad/s. The lowpass transmission function of the third-order elliptic function is then given by

$$S_{21}(s) = \frac{j0.2874s + 1}{0.3935s^3 + (0.9223 - j0.0578)s^2 + (1.3739 - j0.2075)s + (0.9347 - j0.3566)}$$

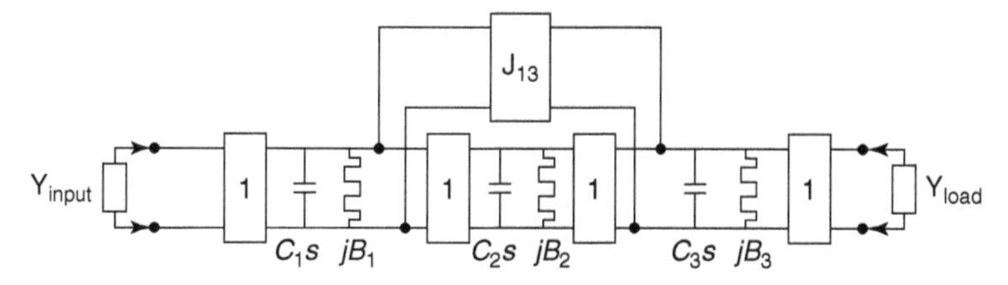

Figure 2.21 Third-order pseudo-elliptic lowpass prototype circuit with one transmission zero.

and the lowpass reflection function is given by

$$S_{11}(s) = \frac{0.3935s^3 - j0.0578s^2 + 0.2930s - j0.0289}{\begin{array}{l}0.3935s^3 + (0.9223 - j0.0578)s^2 \\ \quad +(1.3739 - j0.2075)s + (0.9347 - j0.3566)\end{array}}$$

In the expressions above, scaling was applied so that these functions have a maximum magnitude of 1. For the extraction technique to start properly, the numerator polynomial of S_{11} was multiplied by $-j$. This does not change the magnitude characteristics of the function. Then after extracting the first admittance inverter, the extraction procedure starts with the admittance function.

$$Y(s) = \frac{\begin{array}{l}0.7871s^3 + (0.9223 - j0.1156)s^2 + (1.6669 - j0.2075)s \\ \quad +(0.9347 - j0.3854)\end{array}}{0.9223s^2 + (1.0808 - j0.2075)s + (0.9347 - j0.3277)}$$

The extraction procedure follows the steps provided in Section 2.4.5. The elements of the lowpass prototype are found to be

$$C_1 = 0.853343 \quad C_2 = 1.184017 \quad C_3 = 0.853343$$

$$B_1 = 0.066708 \quad B_2 = -0.358944 \quad B_3 = 0.066708 \quad \text{and} \quad J_{13} = 0.265932$$

One notes that the coupling that creates a transmission zero on the right side of the passband is positive, $J_{13} = 0.265932$.

The third-order bandpass prototype circuit is shown in Figure 2.22. The capacitor and inductor values of the bandpass prototype are given by

$$C_{ai} = \frac{1}{2}\left(\frac{2\alpha C_i + B_i}{\omega_0}\right) \quad \text{and} \quad L_{ai} = \frac{2}{\omega_0(2\alpha C_i - B_i)}$$

In this case we have

$$C_{a1} = 1.6981\text{nF} \quad C_{a2} = 2.3531\text{nF} \quad C_{a3} = 1.6981\text{nF}$$

$$L_{a1} = 0.1113\text{pH} \quad L_{a2} = 0.0801\text{pH} \quad L_{a3} = 0.1113\text{pH} \quad \text{and} \quad J_{13} = 0.265932$$

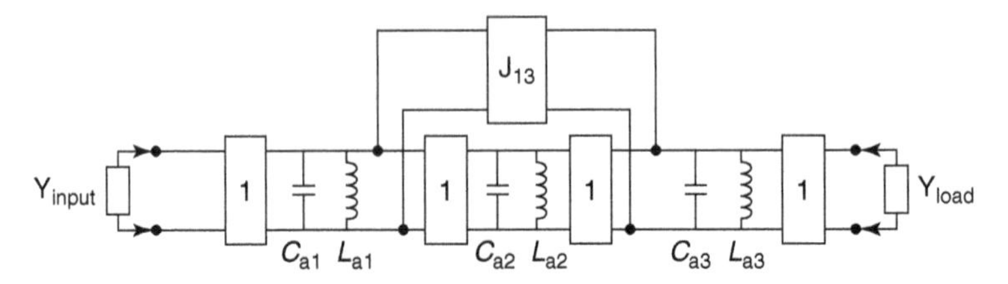

Figure 2.22 Third-order pseudo-elliptic bandpass prototype circuit with one transmission zero.

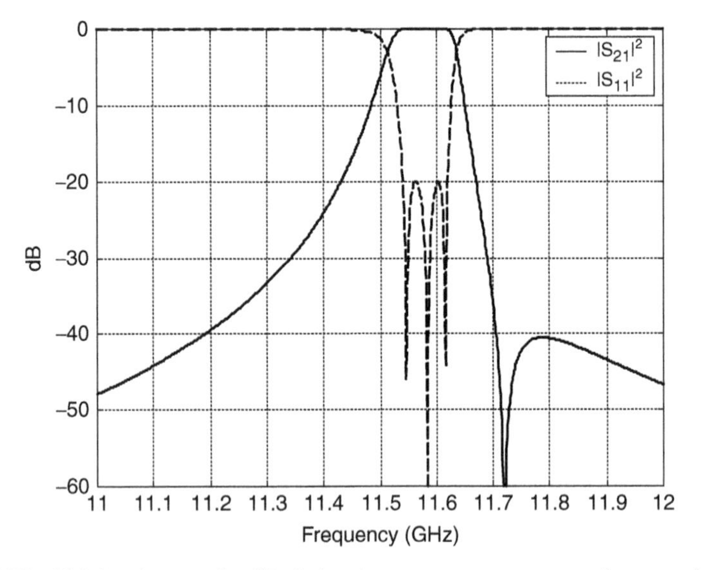

Figure 2.23 Third-order pseudo-elliptic bandpass prototype response (zero on the right).

Simulations of the bandpass prototype circuit of Figure 2.22 provide the bandpass transmission and reflection responses shown in Figure 2.23.

The second example consists of a third-order filter with a transmission zero on the left side of the passband. The only difference from the first example is the location of the zero. This filter has the following specifications:

- *Central frequency:* 11.58 GHz
- *Bandwidth:* 80 MHz
- *Transmission zero frequency:* 11.51 GHz
- *Return loss:* 20 dB

The structure of the lowpass and bandpass prototype circuits will be the same as in the first example. However, in this case, the lowpass transmission zero is such that $\omega_z = -1.7553$ rad/s. In this case it is negative [15,16].

The lowpass transmission function of the third-order elliptic function is then given by

$$S_{21}(s) = \frac{-j0.5697s + 1}{0.3662s^3 + (0.8606 + j0.1145)s^2 + (1.2770 + j0.4029)s + (0.7282 + j0.6878)}$$

and the lowpass reflection function is given by

$$S_{11}(s) = \frac{0.3662s^3 + j0.1145s^2 + 0.2657s + j0.0573}{0.3662s^3 + (0.8606 + j0.1145)s^2 + (1.2770 + j0.4029)s + (0.7282 + j0.6878)}$$

Then after extracting the first admittance inverter, the extraction procedure starts with the admittance function.

$$Y(s) = \frac{0.7324s^3 + (0.8606 + j0.2290)s^3 + (1.5427 + j0.4029)s + (0.7282 + j0.7450)}{0.8606s^2 + (1.0113 + j0.4029)s + (0.7282 + j0.6305)}$$

The extraction procedure follows the steps provided in Section 2.4.5. The elements of the lowpass prototype are found to be

$$C_1 = 0.851033 \quad C_2 = 1.568338 \quad C_3 = 0.851033$$

$$B_1 = -0.132247 \quad B_2 = 0.977848 \quad B_3 = -0.132247 \quad \text{and} \quad J_{13} = -0.563352$$

One notes that the coupling that creates a transmission zero on the right side of the passband is positive, $J_{13} = -0.563352$.

The elements of the bandpass prototype are given by

$$C_{a1} = 1.6922 \text{ nF} \quad C_{a2} = 3.1268 \text{ nF} \quad C_{a3} = 1.6922 \text{ nF}$$

$$L_{a1} = 0.1115 \text{ pH} \quad L_{a2} = 0.0607 \text{ pH} \quad L_{a3} = 0.1115 \text{ pH and } J_{13} = -0.563352$$

Simulations of the bandpass prototype circuit of Figure 2.22 provide the bandpass transmission and reflection responses shown in Figure 2.24.

The third example consists of a third-order filter with three transmission zeros. This filter has the following specifications:

- *Central frequency:* 18.9 GHz
- *Bandwidth:* 200 MHz
- *Transmission zero frequencies:* 18.7, 19.05, and 19.13 GHz
- *Return loss:* 18 dB

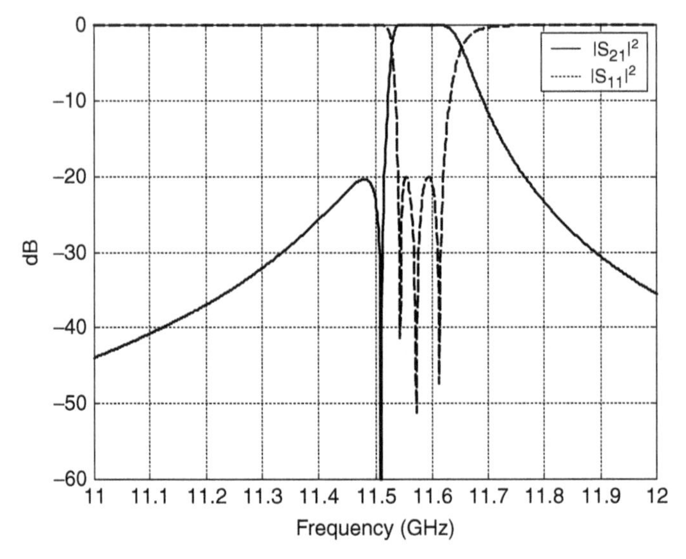

Figure 2.24 Third-order pseudo-elliptic bandpass prototype response (zero on the left).

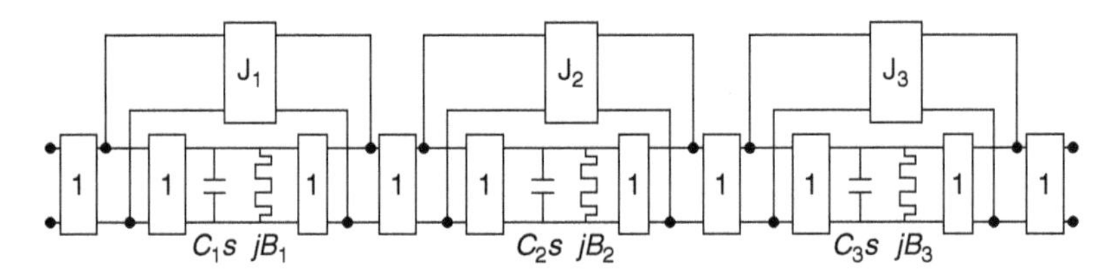

Figure 2.25 Third-order pseudo-elliptic lowpass prototype circuit with three transmission zeros.

The lowpass prototype circuit for this filter is shown in Figure 2.25. The three $C_i - B_i$ resonators will provide the third order (three reflection zeros), and the three nonunitary inverters will provide the three transmission zeros [17]. The first step in the synthesis consists of replacing the fundamental elements by their equivalent circuit. Figure 2.26 provides the equivalent-circuit representation of the prototype circuit to be used for element extraction. The input admittance of the equivalent circuit of Figure 2.25 is given by

$$Y(s) = jJ_1 + \cfrac{1}{\frac{j}{J_1} + \frac{1}{C_1'(s-j\omega_1)} + \cfrac{1}{jJ_1 + \cfrac{1}{jJ_2 + \cfrac{1}{\frac{j}{J_2} + \frac{1}{C_2'(s-j\omega_2)} + \cfrac{1}{jJ_2 + \cfrac{1}{jJ_3 + \cfrac{1}{\frac{j}{J_3} + \frac{1}{C_3'(s-j\omega_3)} + \frac{1}{jJ_3 + 1/Y_0}}}}}}}}$$

The extraction starts with defining the value of the first inverter:

$$jJ_1 = Y(s)|_{s \to j\omega_1}$$

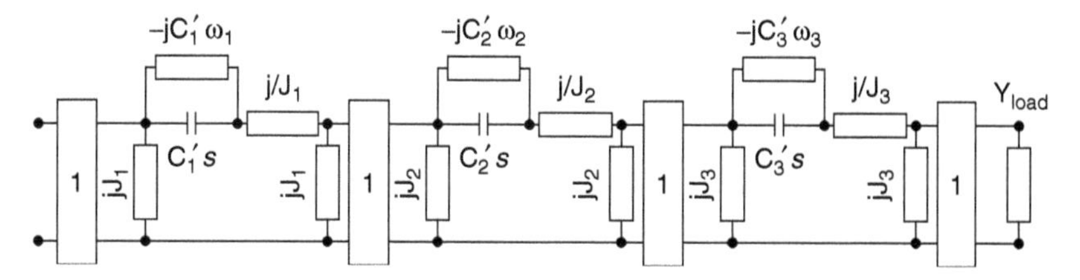

Figure 2.26 Equivalent circuit used for the extraction procedure.

Then we calculate the residue

$$Z_1(s) = \frac{1}{Y(s) - jJ_1} - \frac{j}{J_1} = \frac{1}{C_1'(s - j\omega_1)} + \frac{1}{\cdots}$$

From this we can then extract the first capacitance,

$$\frac{1}{C_1'} = (s - j\omega_1)Z_1(s)|_{s\to j\omega_1}$$

Then we compute new residues such that

$$Z_2(s) = Z_1(s) - \frac{1}{C_1'(s - j\omega_1)s} = \cfrac{1}{jJ_1 + \cfrac{1}{jJ_2 + 1/(j/J_2) + \cdots}}$$

$$Z_3(s) = \frac{1}{Y_2(s) - jJ_1} = jJ_2 + \cfrac{1}{\cfrac{j}{J_2} + \cfrac{1}{C_2'(s - j\omega_2)s} + \cdots}$$

We can then extract the second inverter:

$$jJ_2 = Z_3(s)|_{s\to j\omega_2}$$

Using the same technique, we successively extract C_2', J_3, C_3', and the remaining load, Y_o. The prototype elements are then provided from the equivalent-circuit elements using

$$C_i = \frac{C_i'}{J_i^2} \quad \text{and} \quad B_i = -\frac{J_i + C_i'\omega_i}{J_i^2}$$

The bandpass prototype circuit is given in Figure 2.27. The bandpass prototype element values are then found using the same techniques as in the previous examples:

$$C_{a1} = 1.369 \text{ nF} \qquad C_{a2} = 1.309 \text{ nF} \qquad C_{a3} = 2.360 \text{ nF}$$

$$L_{a1} = 52.32 \text{ fH} \qquad L_{a2} = 53.69 \text{ fH} \qquad L_{a3} = 29.80 \text{ fH}$$

$$J_1 = -0.5780 \qquad J_2 = 0.8696 \qquad J_3 = 0.1976$$

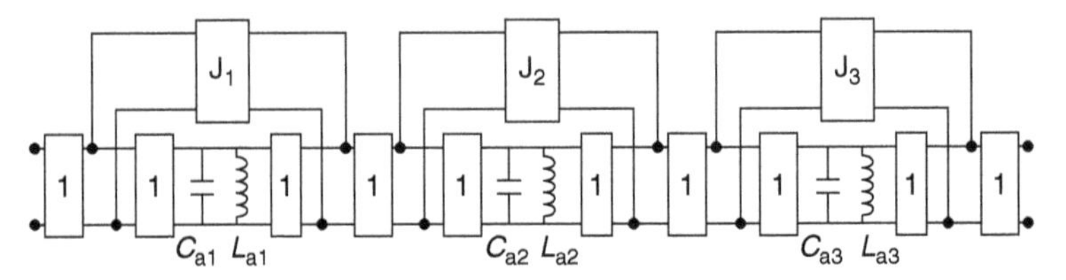

Figure 2.27 Third-order pseudo-elliptic bandpass prototype circuit with three transmission zeros.

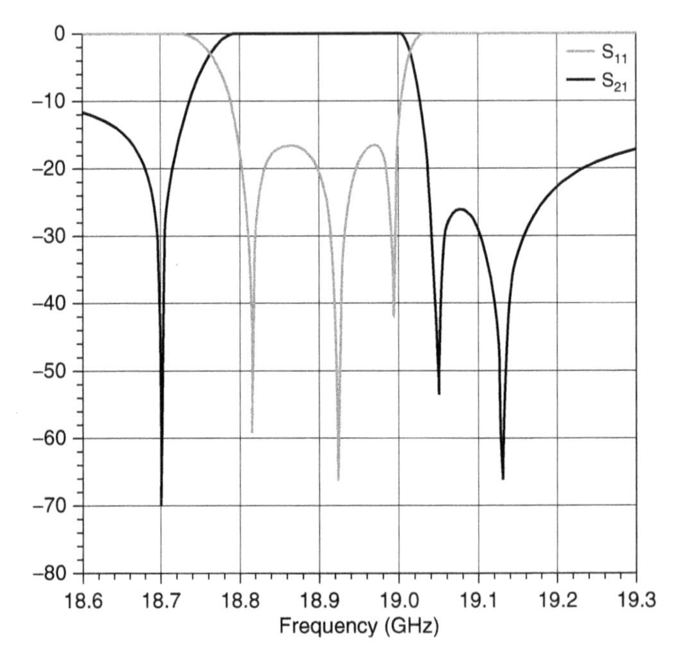

Figure 2.28 Third-order pseudo-elliptic bandpass prototype response with three transmission zeros.

Simulations of the bandpass prototype provide the bandpass transmission and reflection responses shown in Figure 2.28.

The fourth example consists of a fourth-order filter with four transmission zeros. This filter has the following specifications:

- *Central frequency:* 2 GHz
- *Bandwidth:* 50 MHz
- *Transmission zero frequencies:* 1.92, 195, 2.05, and 2.07 GHz
- *Return loss:* 18 dB

The lowpass prototype circuit for this filter is shown in Figure 2.29. The four $C_i - B_i$ resonators will provide the fourth-order (four reflection zeros), and

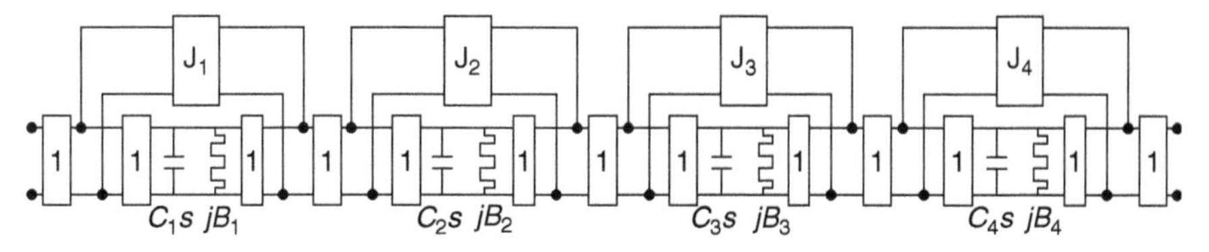

Figure 2.29 Fourth-order pseudo-elliptic lowpass prototype circuit with four transmission zeros.

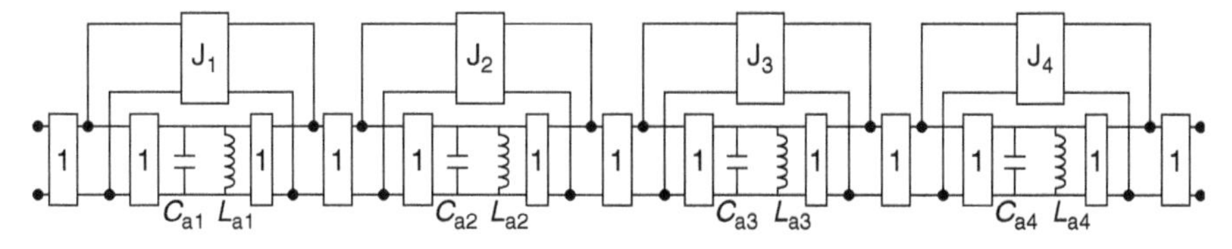

Figure 2.30 Fourth-order pseudo-elliptic bandpass prototype circuit with four transmission zeros.

the four nonunitary inverters will provide the four transmission zeros [14,17,18]. Figure 2.30 gives the bandpass prototype circuit for a fourth-order pseudo-elliptic filter with four transmission zeros. The bandpass prototype element values are then found using the same techniques as in the previous examples.

$$
\begin{array}{llll}
C_{a1} = 4.49 \text{ nF} & C_{a2} = 6.32 \text{ nF} & C_{a3} = 6.70 \text{ nF} & C_{a4} = 2.97 \text{ nF} \\
L_{a1} = 1.43 \text{ pH} & L_{a2} = 0.984 \text{ pH} & L_{a3} = 0.954 \text{ pH} & L_{a4} = 2.10 \text{ pH} \\
J_1 = -0.5405 & J_2 = 0.3759 & J_3 = -0.1730 & J_4 = 0.4587
\end{array}
$$

Simulations of the bandpass prototype provide the bandpass transmission and reflection responses shown in Figure 2.31 when using a final load of 1.28 Ω.

2.6 THEORETICAL COUPLING COEFFICIENTS AND EXTERNAL QUALITY FACTORS

Once the elements of the prototype elements have been computed, the theoretical external quality factors and coupling coefficients can be computed. We use the example of the third-order filter with one transmission zero shown in Figure 2.32. In this case the theoretical external quality factors are given by [2,3]

$$
Q_{\text{input}} = \frac{1}{Y_{\text{input}}} \frac{\omega_{01}}{\omega_0} \left(\frac{C_1}{\beta} + \frac{B_1}{2} \right)
$$

$$
Q_{\text{output}} = \frac{1}{Y_{\text{load}}} \frac{\omega_{03}}{\omega_0} \left(\frac{C_3}{\beta} + \frac{B_3}{2} \right)
$$

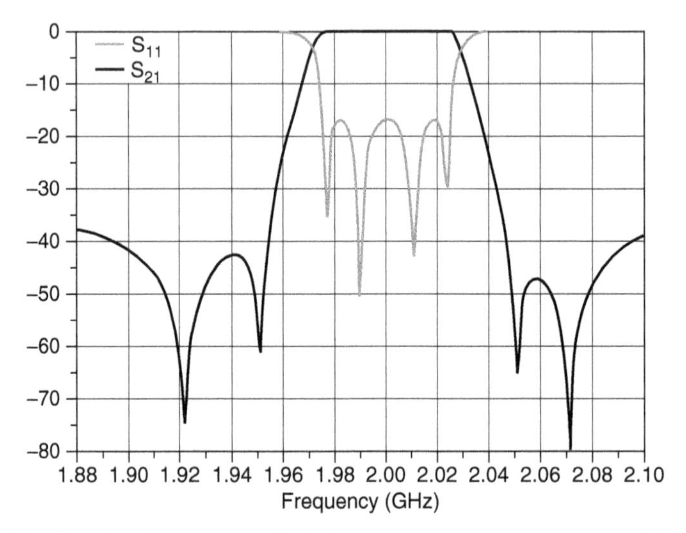

Figure 2.31 Fourth-order pseudo-elliptic bandpass prototype response with four transmission zeros.

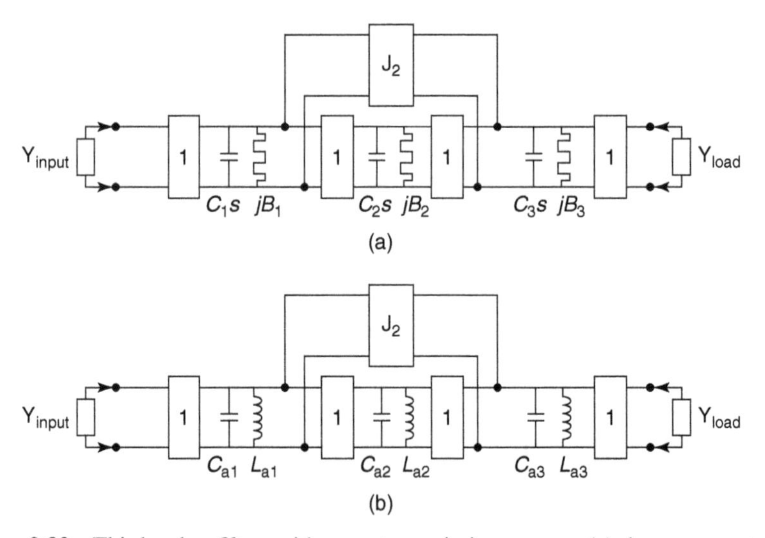

Figure 2.32 Third-order filter with one transmission zero: (a) lowpass prototype; (b) bandpass prototype.

where $\beta = \Delta f / f_0$ is the fractional bandwidth and where the resonant frequencies ω_{0i} of the resonators for the asymmetrical case can be slightly different from the resonant frequency ω_0 of the filter:

$$\omega_{0i} = \frac{1}{\sqrt{L_{ai} C_{ai}}} = \omega_0 \sqrt{1 - \frac{B_i}{C_i \,/\, \beta + B_i \,/\, 2}}$$

In this case the theoretical couplings coefficients are given by

$$K_{ij} = \frac{\omega_0}{\sqrt{\omega_{0i}\omega_{0j}}} \frac{\beta J_{ij}}{\sqrt{(C_i + \beta B_i / 2)(C_j + \beta B_j / 2)}} \quad \text{for } i \neq j$$

As shown in later chapters, the physical parameters of the fractal filters are defined by matching experimental couplings and external quality factors with the theoretical values provided by the prototype.

REFERENCES

[1] G. L. Matthaei, L. Young, and E. M. T. Jones, *Microwave Filters, Impedance-Matching Networks, and Coupling Structures*, Artech House, Norwood, MA, 1985.

[2] J. S. Hong and M. J. Lancaster, *Microstrip Filters for RF/Microwave Applications*, Wiley-Interscience, Hoboken, NJ, 2001.

[3] P. Jarry and J. Beneat, *Advanced Design Techniques and Realizations of Microwave and RF Filters*, Wiley–IEEE Press, Hoboken, NJ, 2008.

[4] R. J. Cameron, Fast generation of Chebyshev filter prototypes with asymmetrically prescribed transmission zeros, *ESA Journal*, Vol. 6, pp. 83–95, 1982.

[5] C. Guichaoua, P. Jarry, and C. Boschet, Approximation, synthesis and realization of pseudo-elliptic asymmetrical microwave filters, *Technical Note CNES 124*, DCAF02583, pp. 188–192, 1990.

[6] R. J. Cameron, General coupling matrix synthesis methods for Chebyshev filtering functions, *IEEE Transactions on Microwave Theory and Techniques*, Vol. 47, No. 4, pp. 433–442, Apr. 1999.

[7] S. A. Mohamed and L. F. Lind, Construction of lowpass filter transfer functions having prescribed phase and amplitude characteristics, *IEEE Colloquium Digest: Microwave Filters*, pp. 8/1–8/5, Jan. 1982.

[8] O. Roquebrun, P. Jarry, E. Kerhervé, and M. Guglielmi, Inline microwave filters with transmission zeros, *ESA Final Report of the Contract*, XRM/069.97/ MG, Apr. 1998.

[9] E. Hanna, P. Jarry, E. Kerhervé, and J. M. Pham, General prototype network method for suspended substrate microwave filters with asymmetrical prescribed transmission zeros: synthesis and realization, *Proceedings of the Mediterranean Microwave Symposium* (*MMS 2004*), Marseille, France, June 2004.

[10] P. Jarry, M. Guglielmi, J. M. Pham, E. Kerhervé, O. Roquebrun, and D. Schmitt, Synthesis of dual-modes in-line asymmetric microwave rectangular filters with higher modes, *International Journal of RF and Microwave Computer-Aided Engineering*, pp. 241–248, Feb. 2005.

[11] D. S. G. Chambers and J. D. Rhodes, A low-pass prototype network allowing the placing of integrated poles at real frequencies, *IEEE Transactions on Microwave Theory and Techniques*, Vol. 31, No. 1, pp. 40–45, Jan. 1983.

[12] J. D. Rhodes and S. A. Alseyab, The generalized Chebyshev low-pass prototype filter, *International Journal of Circuit Theory and Applications*, Vol. 8, pp. 113–125, 1980.

[13] R. M. Kurzrok, General three-resonator filters in waveguide, *IEEE Transactions on Microwave Theory and Techniques*, Vol. 14, No. 11, pp. 46–47, Jan. 1966.

[14] E. Hanna, P. Jarry, J. M. Pham, and E. Kerhervé, A design approach for capacitive gap-parallel-coupled-lines filters with the suspended substrate technology, in *Microwave Filters and Amplifiers* (scientific editor P. Jarry), Research Signpost, Tivandrum, India, pp. 49–71, 2005.

[15] E. Hanna, P. Jarry, E. Kerhervé, and J. M. Pham, Suspended substrate stripline cross-coupled trisection filters with asymmetrical frequency characteristics, *IEEE 35th European Microwave Conference (EuMC 2005)*, Paris, pp. 273–276, Oct. 2005.

[16] E. Hanna, P. Jarry, E. Kerhervé, and J. M. Pham, Cross-coupled suspended stripline trisection bandpass filters with open-loop resonators, *IEEE International Microwave and Optoelectronics Conference*, Brasília, Brazil, pp. 42–46, July 25–28, 2005.

[17] E. Hanna, P. Jarry, E. Kerhervé, and J. M. Pham, General prototype network method for suspended substrate microwave filters with asymmetrically prescribed transmission zeros: synthesis and realization, *IEEE 2004 Mediterranean Microwave Symposium*, Marseille, France, pp. 40–43, June 1–3, 2004.

[18] E. Hanna, P. Jarry, E. Kerhervé, and J. M. Pham, General prototype network with cross coupling synthesis method for microwave filters: application to suspended substrate strip line filters, *ESA–CNES International Workshop on Microwave Filters*, Toulouse, France, Sept. 13–15, 2004.

CHAPTER 3

SUSPENDED SUBSTRATE STRUCTURE

3.1 Introduction 63
3.2 Suspended substrate technology 64
3.3 Unloaded quality factor of a suspended substrate resonator 66
3.4 Coupling coefficients of suspended substrate resonators 69
3.4.1 Input/output coupling 69
3.4.2 Coupling between resonators 69
3.5 Enclosure design considerations 73
References 75

3.1 INTRODUCTION

The fractal filters in this book were fabricated using suspended substrate technology. This section gives the necessary background to provide the reader with a better understanding of the fabrication details necessary for completing an actual microwave filter.

3.2 SUSPENDED SUBSTRATE TECHNOLOGY

The use of printed transmission lines for microwave circuits has increased considerably for several reasons: They are simple to fabricate, they are operational on a large bandwidth, and they have small dimensions. This makes them very useful for compact filtering structures.

A first type of printed line is the stripline [Figure 3.1.(a)]. It consists of a conducting line centered between two ground planes and two identical substrates. A small airgap can exist between the two substrates because of fabrication defaults, thus degrading the propagation performances. To enhance stripline performances, configuration (b) is used: The coupling between the two conductors reduces the electric field in the airgap.

The suspended stripline is a modified version of the stripline: The conducting line is printed on a thin substrate suspended in a metallic box. Most of the electromagnetic field is confined between the airgaps and the ground planes [1,2]. Configuration (c) is the most popular in suspended substrate technology and is used in this book to implement fractal filters. Configuration (d) of suspended

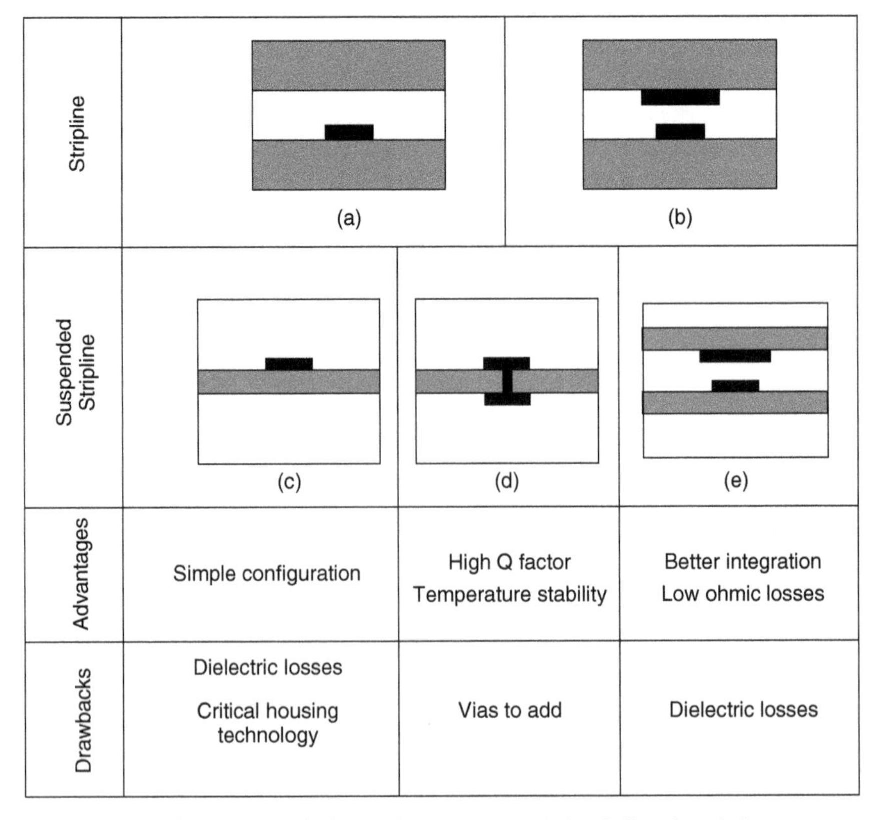

Figure 3.1 (a,b) Stripline and (c–e) suspended stripline descriptions.

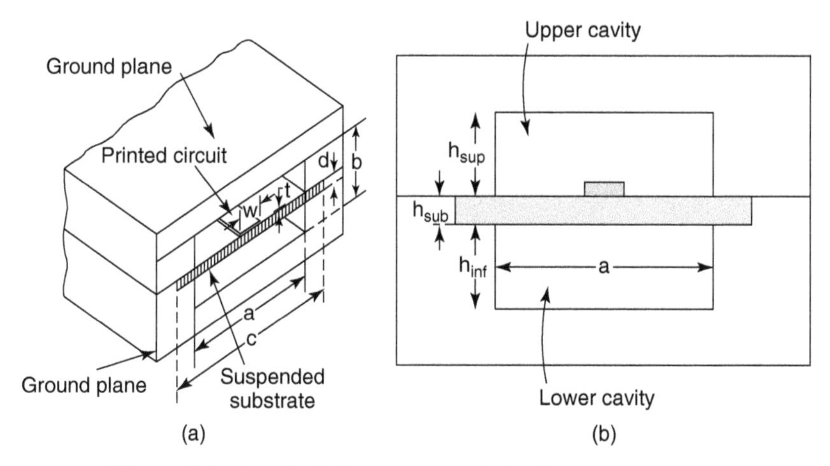

Figure 3.2 Detailed view of suspended substrate technology.

stripline shows that two strips are printed symmetrically on both sides of the substrate. In this case strong coupling is achieved and parasitic propagation modes are avoided. However, this configuration requires high-precision fabrication tolerances. Figure 3.2 shows in more detail how the suspended substrate stripline of configuration (c) is realized.

The performance of suspended substrate depends considerably on the choice of the substrate. A thin substrate as well as a small relative permittivity ($1 < \varepsilon_r < 3.8$) provides the following advantages over conventional microstrip [3]:

- Higher quality factors (500) are obtainable as well as lower losses, since most of the energy is propagated through the airgaps.
- Easier fabrication for millimeter-wave applications since air as a propagation medium makes it possible to increase the global circuit dimensions.
- A wider range of achievable impedance values (150 Ω).
- Operation over a wider frequency range.
- No electromagnetic radiation since the circuit is enclosed in a box.
- Can be combined with other planar structures (microstrip, coplanar, slotline).
- More stable to temperature than microstrip or stripline filters.
- Greater resistance to environment vibrations, which explains why they are used so widely in military applications.

The drawbacks of suspended substrate technology are difficulty in miniaturization (because of the fabrication tolerances) and the enclosure design.

Principal references regarding suspended substrate technology, by category, are as follows: generalities about suspended substrate technology and its principal characteristics [1–3], characterization, synthesis equations, and impedance calculations [4–18], study of discontinuities (couplings, line-width variations)

[19–24], propagation characteristics [25–30], and examples of circuits using suspended substrate technology [31–53].

3.3 UNLOADED QUALITY FACTOR OF A SUSPENDED SUBSTRATE RESONATOR

The unloaded quality factor Q_0 of a resonator provides a measure of its electrical performance. It gives an idea of the insertion loss of the filter and the out-of-band rejection (selectivity of the filter) to be expected. The unloaded quality factor is defined as follows:

$$Q_0 = \omega_0 \frac{\text{energy stored in the resonator}}{\text{energy lost by radiation}}$$

where ω_0 is the resonance frequency of the resonator. At resonance, this expression becomes

$$Q_0 = \omega_0 \frac{\overline{W}_m + \overline{W}_e}{\overline{P}} = \omega_0 \frac{2\overline{W}_m}{\overline{P}} = \omega_0 \frac{2\overline{W}_e}{\overline{P}}$$

where $\overline{W}_m$ and $\overline{W}_e$ are, respectively, the average magnetic and electric energy stored in the resonator. At resonance, these two energies are equal. $\overline{P}$ represents the ohmic losses, the dielectric losses, and the radiation losses.

The unloaded quality factor can also be determined using an equivalent electrical circuit. Two low-frequency models are often used: the series configuration and the parallel configuration (Figure 3.3). For these cases, the quality factor is given by

$$Q_0 = \begin{cases} \dfrac{L\omega}{R} & \text{(series configuration)} \\[2ex] \dfrac{R}{L\omega} & \text{(parallel configuration)} \end{cases}$$

with $\omega = 1/\sqrt{LC}$. In order to characterize suspended substrate technology, the behavior of the unloaded quality factor of a half-wavelength resonator, together

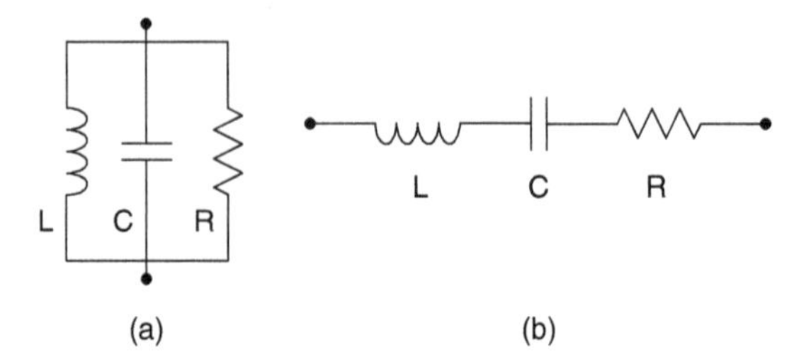

Figure 3.3 Low-frequency resonator model: (a) parallel configuration; (b) series configuration.

with the variations of the geometrical parameters of the suspended substrate (i.e., the height of the upper and lower cavities), have been examined by electromagnetic analysis using Agilent's Momentum software. The unloaded quality factor can be defined from the transmission characteristic of a weakly coupled resonator. The input/output couplings are kept weak so that they don't interfere with the couplings inside the resonator, affecting the quality factor. It was found that input/output couplings do not affect the unloaded quality factor of the resonator when at least 20 dB of attenuation is achieved at the maximum of the transmission characteristic of the resonator (Figure 3.4).

The loaded quality factor of the resonator is then defined by

$$Q_L = \frac{f_0}{\Delta f}$$

where f_0 is the resonance frequency of the resonator and Δf is the -3 dB bandwidth. The unloaded quality factor is then computed from the loaded quality factor using

$$Q_0 = \frac{Q_L}{1 - S_{21}(f_0)}$$

where $S_{21}(f_0)$ is the amplitude (in linear scale) of the transmission characteristic at the resonance frequency.

The change in the unloaded quality factor of a 10-GHz resonator when the cavity height (either the upper or lower height) is varied is examined next. In this case the substrate was Duroid 5880 with a relative permittivity of 2.2 (loss

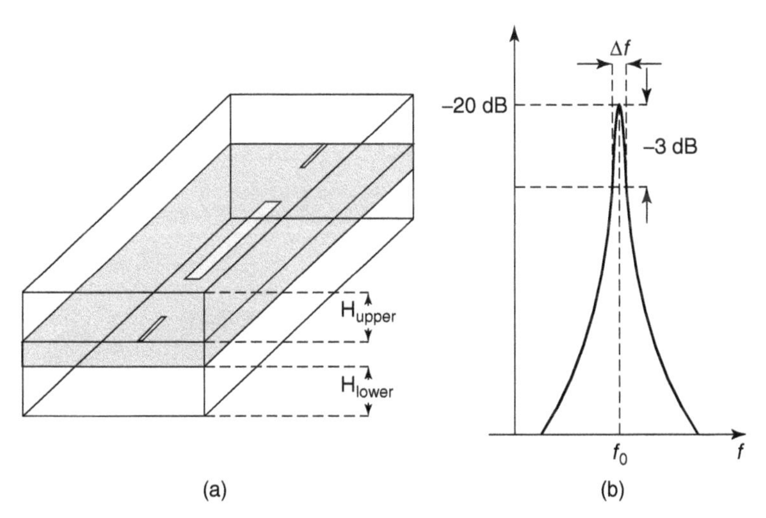

Figure 3.4 Calculation of the unloaded quality factor of a suspended substrate resonator: (a) topology of the weakly coupled resonator; (b) calculation using the magnitude of S_{21}.

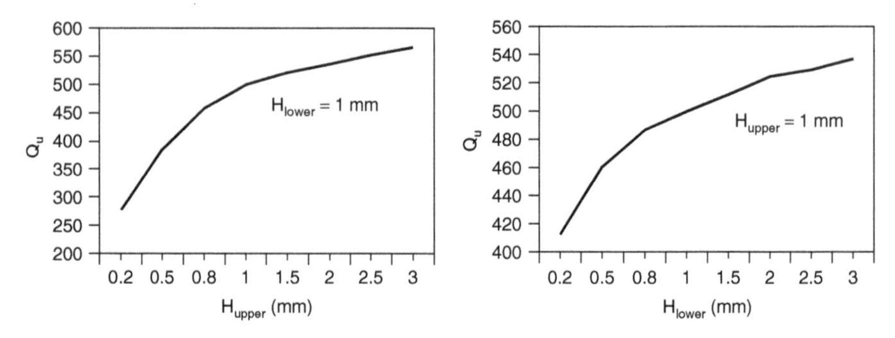

Figure 3.5 Unloaded quality factor of a resonator as a function of cavity height.

tangent = 0.0009) and a thickness of 0.254 mm. In the results the line width is kept constant while its length is adjusted to keep the resonance at 10 GHz when varying the cavity height. When varying the upper height, the lower height is kept fixed at 1 mm, and vice versa. Figure 3.5 shows the variations in the unloaded quality factor of the resonator when the upper or lower heights are changed.

The unloaded quality factor increases when either the lower or upper height increases. If we consider the case where the lower height is fixed at 1 mm and the upper height is 0.3 mm, much smaller than the lower height, the electrical field loops back on both the upper and lower ground planes [Figure 3.6(a)]. When the upper height is much greater than the lower height, the electrical field loops back mainly on the lower ground plane [Figure 3.6(b)]. In this case, the metallic losses are therefore smaller and the unloaded quality factor increases.

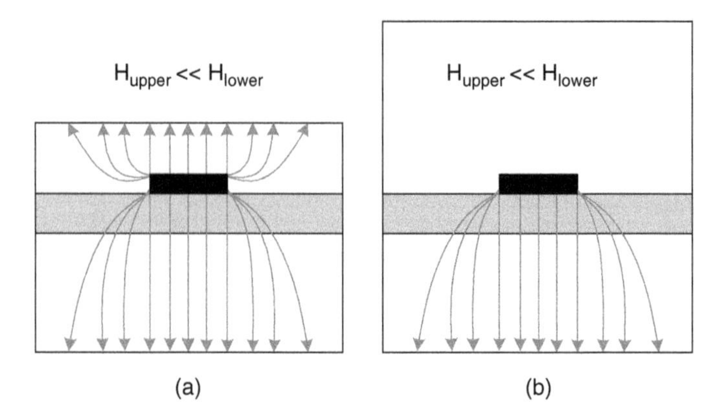

Figure 3.6 Electrical field distribution depending on the upper cavity height.

3.4 COUPLING COEFFICIENTS OF SUSPENDED SUBSTRATE RESONATORS

The next stage is to define the coupling coefficients using suspended substrate for the input and output of the filter as well as the coupling coefficients between resonators. In all of the following applications, the substrate is Duroid 5880 with a relative permittivity of 2.2 and a thickness of 0.254 mm.

3.4.1 Input/Output Coupling

The input coupling characterizes the excitation of the first resonator. The input coupling depends on the distance (g) between the input line and the first resonator, as shown in Figure 3.7(a). The output coupling depends on the distance between the last resonator and the output line. To calculate the input or output coupling, we consider the case of a half-wavelength resonator excited by a 50-Ω impedance line [Figure 3.7(a)]. The input coupling is computed using measures of the reflection coefficient S_{11}. In Figure 3.7(b) f_0 is the central frequency and Δf is the bandwidth corresponding to a 180° phase shift around the center frequency. The external quality factor representing the input coupling is then given by

$$Q_e = \frac{f_0}{\Delta f}$$

3.4.2 Coupling Between Resonators

The coupling between two resonators depends on the gap distance between them. When two resonators are close to each other, their associated resonant frequencies shift due to the coupling. To define the coupling, the setup of Figure 3.8(a) is used.

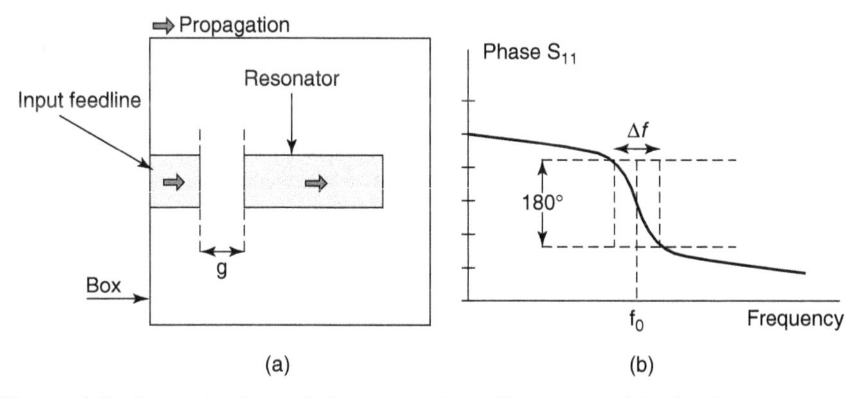

Figure 3.7 Determination of the external quality factor: (a) simulated structure; (b) calculation using the phase of S_{11}.

The input/output couplings are kept weak so that they do not interfere with the coupling between resonators. These couplings are assumed not to interfere with the coupling between resonators when at least 10 dB of attenuation is achieved between the two resonant frequencies f_1 and f_2, as shown in Figure 3.8(b). The coupling coefficient K between resonators is then defined by

$$K = \frac{f_2^2 - f_1^2}{f_2^2 + f_1^2}$$

where f_1 and f_2 are the resonant frequencies of the two weakly coupled resonators.

Variations in resonant frequencies f_1 and f_2 of two 10-GHz half-wavelength resonators when the gap distance changes are shown in Figure 3.9. When the gap g between resonators decreases, f_1 decreases and f_2 increases, so that the coupling coefficient K increases. Figure 3.10 shows variations in the coupling coefficient K as a function of g for two cases of cavity heights. In case A, the

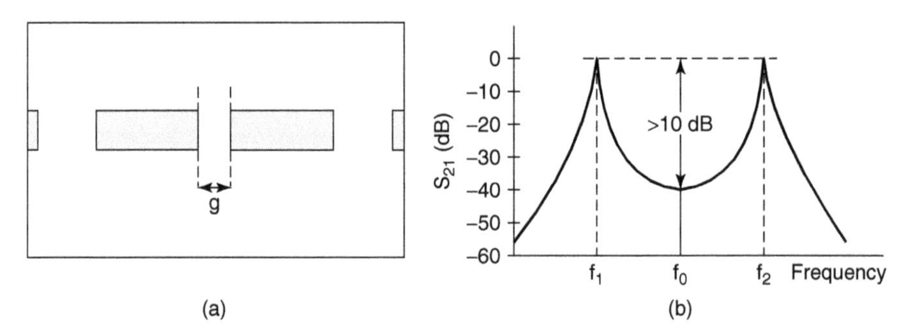

Figure 3.8 Determination of the coupling between resonators: (a) simulated structure; (b) calculation using the magnitude of S_{21}.

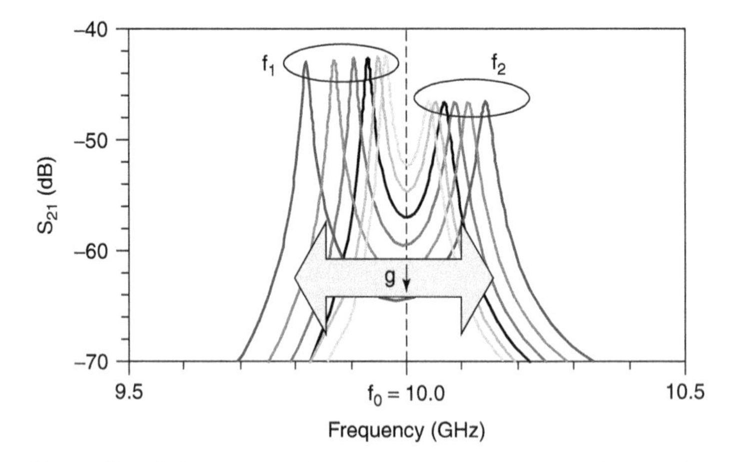

Figure 3.9 Effect of gap distance on resonator resonant frequencies.

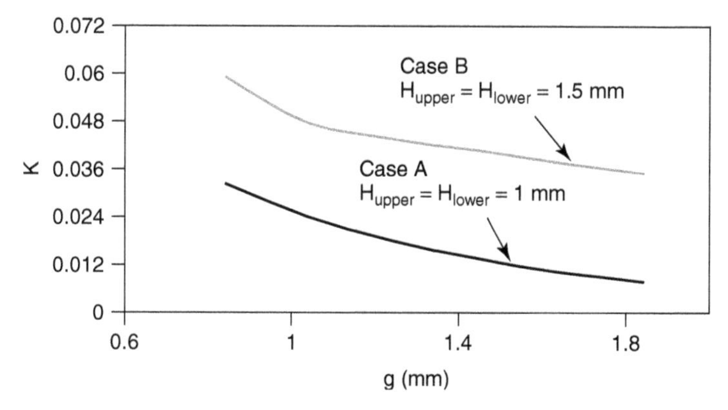

Figure 3.10 Effect of cavity heights on the coupling coefficient K.

upper and lower cavity heights are equal to 1 mm, and in case B, they are equal to 1.5 mm. These simulations show that for the same gap width, the coupling coefficient is stronger in case B (i.e., when the cavity heights increase).

In Chapter 2 we saw that a positive coupling between nonadjacent resonators creates a transmission zero on the right side of the passband, and a negative coupling creates a transmission zero on the left side of the passband. Next, we show how negative and positive couplings can be achieved in practice.

The notion of positive or negative coupling is somewhat relative. When a coupling is considered positive, the negative coupling will have an opposite phase characteristic. For example, using simulations, Figure 3.11 shows

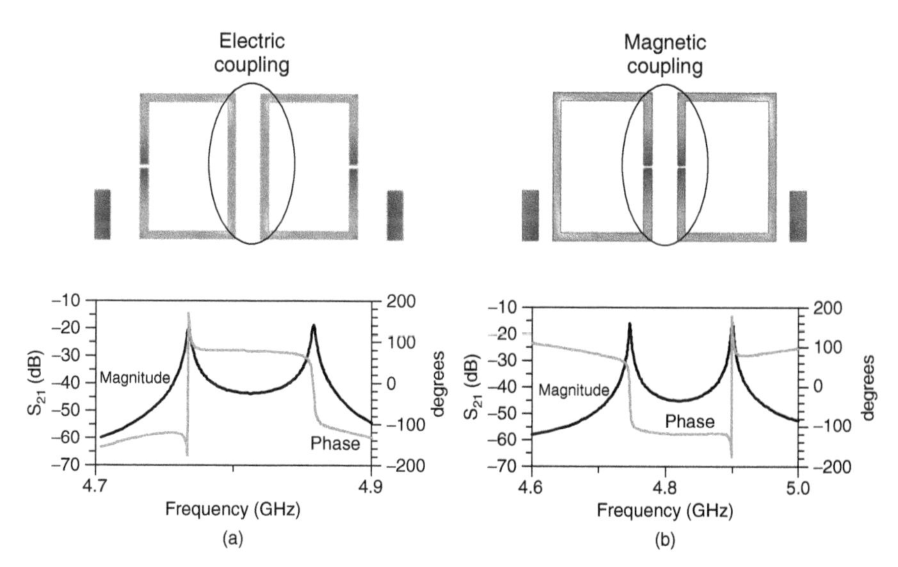

Figure 3.11 (a) Electric coupling between two resonators; (b) magnetic coupling between two resonators.

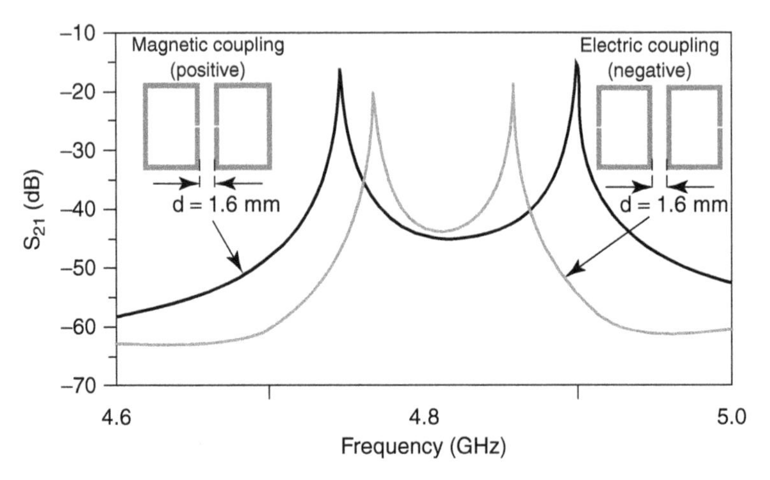

Figure 3.12 Comparison of magnetic and electric coupling.

the amplitude and phase characteristics of two $\lambda/2$ rectangular resonators. In configuration (a) the electric field is concentrated essentially in the coupling zone. In this case the coupling is considered *electric*. In configuration (b) the electric field is concentrated outside the coupling zone. In this case the coupling is considered *magnetic*. One can see that the phase characteristics of the two configurations are opposite, leading to the concept of positive or negative coupling.

It is interesting to note that for the same gap distance the magnetic coupling is stronger than the electric coupling. Figure 3.12 shows the transmission characteristics of two magnetically and electrically coupled resonators. In both cases the gap distance is 1.6 mm. In this case it is found that the magnetic coupling (positive coupling) is $K = 0.0317$ and the electric coupling (negative coupling) is $K = 0.0187$.

Depending on the application (narrowband or wideband filter) and depending on the desired position of the transmission zeros, electric or magnetic coupling can be preferred to achieve the transfer function requested. For example, by combining positive and negative couplings, transmission zeros can be created. Figure 3.13(a) shows a three-resonator circuit with an electric coupling between resonators 1 and 3. This coupling creates a transmission zero on the left side of the resonances (which can be thought of as the passband of a filter). Figure 3.13 (b) shows a three-resonator circuit with a magnetic coupling between resonators 1 and 3. In this case the transmission zero appears on the right side of the resonances. Comparing these results to those of the synthesis examples of Chapter 2, we conclude that the electric coupling will represent a *negative coupling* and the magnetic coupling will represent a *positive coupling*.

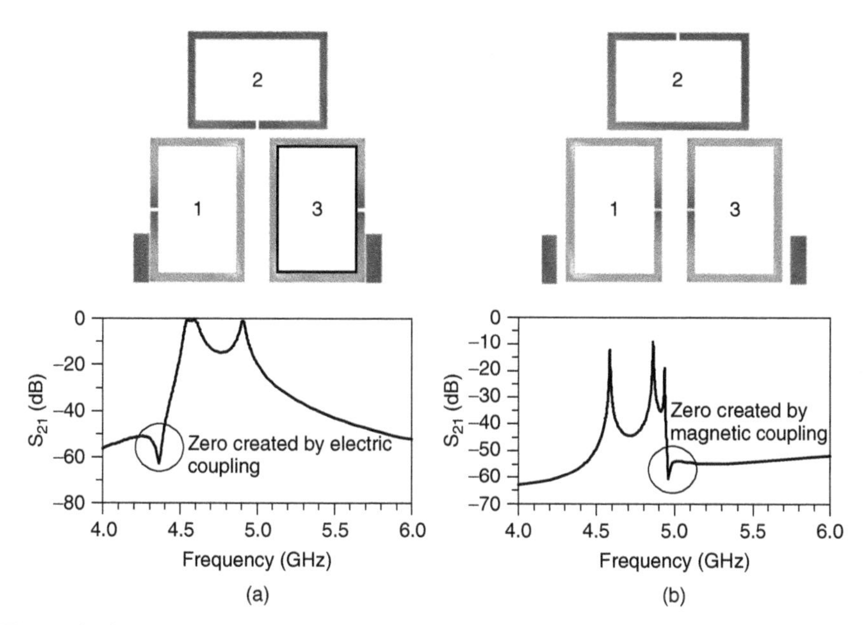

Figure 3.13 Transmission zero created by (a) an electric coupling and (b) a magnetic coupling.

3.5 ENCLOSURE DESIGN CONSIDERATIONS

In suspended substrate structures, the enclosure cannot be neglected. It must be designed carefully to avoid propagation of waveguide modes that can seriously damage the transfer characteristics of the filter by creating unwanted couplings. The enclosure can act as a microwave cavity and have a resonant frequency. Figure 3.14 provides a model of the enclosure with the suspended substrate. The resonant frequency of a given mode of propagation n,m for the enclosure filled with three dielectric materials shown in Figure 3.14 is given by

$$f_{m,n,p} = \frac{c}{2\pi\sqrt{\varepsilon_{\mathrm{eq}}}}\sqrt{\left(\frac{m\pi}{l}\right)^2 + \left(\frac{n\pi}{L}\right)^2 + \left(\frac{p\pi}{h}\right)^2}$$

$$\varepsilon_{\mathrm{eq}} = \frac{h}{h_1/\varepsilon_{r1} + h_2/\varepsilon_{r2} + h_3/\varepsilon_{r3}}$$

where $\varepsilon_{\mathrm{eq}}$ is the equivalent relative permittivity of the cavity; the ε_{ri} are the relative permittivities of the three dielectric materials; L,l and h are the length, width, and height of the cavity, respectively, and m,n, and p are the indexes of the propagation modes. For our filters, we will have $\varepsilon_{r1} = \varepsilon_{r3} = 1$ (air) and $\varepsilon_{r2} = 2.2$ (Duroid 5880).

When designing the enclosure, the dimensions should be defined carefully so that the propagation modes do not resonate near the operational passband of the filter. By letting the length/width ratio $R = L/l$, the resonance frequency of the

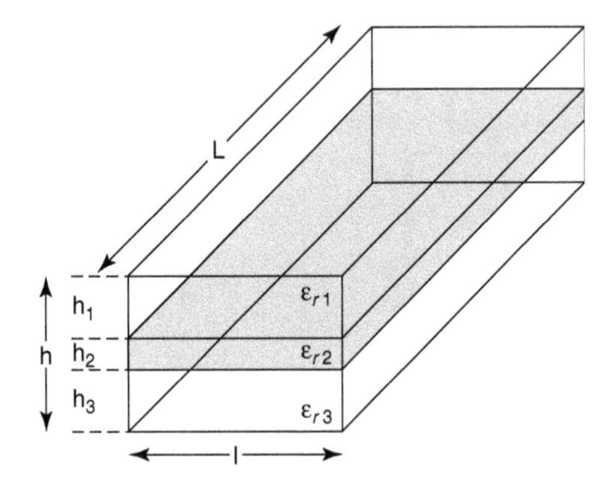

Figure 3.14 Waveguide structure loaded with a substrate.

TE_{mn0} modes can be expressed as

$$f_{mn0} = \frac{c}{2L\sqrt{\varepsilon_{\rm eq}}}\sqrt{n^2 + m^2R^2}$$

The length L of the cavity is generally estimated based on the topology and order of the filter and on some preliminary values from the filter design technique. Once the length of the cavity is defined, the resonant frequency of the different modes can be computed as a function of the length/width ratio R.

Figure 3.15 shows the resonant frequencies for a few modes as a function of the length/width ratio R for a filter that would have a center frequency of 11 GHz. The

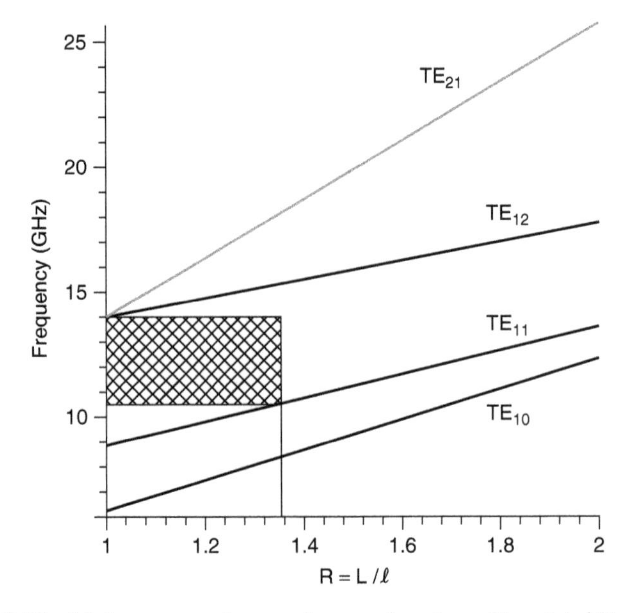

Figure 3.15 Mode resonant frequencies as a function of length/width ratio R.

cavity length was estimated at $L = 23.2$ mm, and the various heights were $h_1 = h_3 = 1$ mm and $h_2 = 0.254$ mm. We observe that in the shaded zone there are no mode resonant frequencies. In addition, this zone corresponds to frequencies from 10.5 to 14.5 GHz, a good representative of the operational band of a filter with a center frequency of 11 GHz. Therefore, one should select R to fit in this zone (R between 1 and 1.35). After selecting the value of R, it is possible to compute the width l of the cavity. Once L,l, and h are defined, we can check more carefully the distribution of the mode resonant frequencies for the cavity.

REFERENCES

[1] L. G. Maloratsky, Reviewing the basics of suspended striplines, *Microwave Journal*, Oct. 2002.

[2] J. E. Dean, Suspended substrate stripline filters for ESM applications, *IEE Proceedings*, Vol. 132, No. 4, pp. 257–266, July 1985.

[3] J. D. Rhodes, Suspended substrate provides alternative to coax, *MSN Journal*, pp. 134–143, Aug. 1979.

[4] J. Smith, The even and odd mode capacitance parameters for coupled lines in suspended substrate, *IEEE Transactions on Microwave Theory and Techniques*, Vol. 19, No. 5, pp. 424–431, May 1971.

[5] P. Bhartia and P. Pramanick, Computer-aided design models for broadside-coupled striplines and millimeter-wave suspended substrate microstrip lines, *IEEE Transactions on Microwave Theory and Techniques*, Vol. 36, No. 11, pp. 1476–1481, Nov. 1988.

[6] R. Tomar, Y. Antar, and P. Bhartia, CAD of suspended substrate microstrips: an overview, *International Journal of RF and Microwave Computer-Aided Engineering*, Vol. 15, No. 1, pp. 44–55, Jan. 2005.

[7] L. Jingshun, Computer-aided design of elliptic function suspended substrate filters, *Proceedings of the International Conference on Microwave and Millimeter Wave Technology (ICMMT)*, Beijing, China, pp. 917–920, 1998.

[8] E. Yamashita and K. Atsuki, Strip line with rectangular outer conductor and three dielectric layers, *IEEE Transactions on Microwave Theory and Techniques*, Vol. 18, No. 5, pp. 238–244, May 1970.

[9] Y. Wang, K. Gu, and Y. Shu, Analysis equations for shielded suspended substrate microstrip line and broadside-coupled structure, *IEEE Transactions on Microwave Theory and Techniques*, Vol. 2, pp. 693–696, 1987.

[10] Y. Wang, K. Gu, and Y. Shu, Synthesis equations for shielded suspended substrate microstrip line and broadside-coupled structure, *IEEE Transactions on Microwave Theory and Techniques*, Vol. 4, pp. 452–455, 1988.

[11] Y. Wang, K. Gu, and Y. Shu, Analysis and synthesis equations for edge-coupled suspended substrate microstrip lines, *IEEE Transactions on Microwave Theory and Techniques*, Vol. 3, pp. 1123–1126, 1989.

[12] E. Yamashita, K. Atsuki, and M. Nakajima, Analysis method for generalized suspended striplines, *IEEE Transactions on Microwave Theory and Techniques*, Vol. 34, No. 12, pp. 261–264, Dec. 1986.

[13] A. Lehtuvuori and L. Costa, Model for shielded suspended substrate microstrip line, Helsinki University of Technology, Department of Electrical and Communications Engineering, Circuit Theory Laboratory, Rapport Interne 1998, http://www.aplac.hut.fi/publications/ct-38/ct-38.pdf.

[14] M. Nakajima and E. Yamashita, Characterization of coupled asymmetric suspended striplines having three thick-strip conductors and side-wall grooves, *IEEE International Microwave Symposium Digest*, Vol. 2, pp. 719–722, 1989.

[15] I. P. Polichronakis and S. S. Kouris, Higher order modes in suspended substrate microstrip lines, *IEEE Transactions on Microwave Theory and Techniques*, Vol. 41, No. 8, pp. 1449–1454, Aug. 1993.

[16] R. Kumar, Design model for broadside-coupled suspended substrate stripline for microwave and millimeter-wave applications, *Microwave and Optical Technology Letters*, Vol. 42, No. 4, pp. 328–331, Aug. 2004.

[17] R. Schwindt and C. Nguyen, Analysis of three multilayer parallel-coupled lines, *IEEE Transactions on Microwave Theory and Techniques*, Vol. 42, pp. 1429–1434, July 1994.

[18] L. Zhu and E. Yamashita, Full-wave boundary integral equation method for suspended planar transmission lines with pedestals and finite metallization thickness, *IEEE Transactions on Microwave Theory and Techniques*, Vol. 41, No. 3, pp. 478–483, Mar. 1993.

[19] C. Amrani, M. Drissi, V. F. Hanna, and J. Citerne, Theoretical and experimental investigation of some general suspended substrate discontinuities, *IEEE International Microwave Symposium*, Vol. 1, pp. 409–412, 1992.

[20] L. K. Wu and H. M. Chang, Analysis of dispersion and series gap discontinuity in shielded suspended striplines with substrate mounting grooves, *IEEE Transactions on Microwave Theory and Techniques*, Vol. 40, No. 2, pp. 279–284, Feb. 1992.

[21] R. Sabry and S. K. Chaudhuri, Analysis of suspended stripline discontinuities using full-wave spectral domain approach, *IEEE International Microwave Symposium*, Chicago, pp. 2288–2291, July 18–25, 1992.

[22] Y. S. Xu and Y. Wang, Analysis of discontinuities of microstrip and suspended substrate lines, *IEEE International Microwave Symposium Digest*, Vol. 1, pp. 413–416, 1992.

[23] A. Biswas and V. K. Tripathi, Modeling of asymmetric and offset gaps in shielded microstrip and slotline, *IEEE Transactions on Microwave Theory and Techniques*, Vol. 38, No. 6, pp. 818–822, June 1990.

[24] A. Rong, Transmission characteristics of suspended stripline with finite metallization thickness and substrate supporting grooves, *Electronics Letters*, Vol. 26, pp. 431–432, 1990.

[25] H. Mizuno, C. Verver, R. Douville, and M. G. Stubbs, Propagation in broadside-coupled suspended-substrate stripline in E-plane, *IEEE Transactions on Microwave Theory and Techniques*, Vol. 33, No. 10, pp. 946–950, Oct. 1985.

[26] S. B. Cohn and G. D. Osterhues, A more accurate model of the TE_{10} type waveguide mode in suspended substrate, *IEEE Transactions on Microwave Theory and Techniques*, Vol. 30, No. 3, pp. 293–294, Mar. 1982.

[27] R. S. Tomar and P. Bhartia, Modeling the dispersion in a suspended microstrip line, *IEEE Transactions on Microwave Theory and Techniques*, Vol. 2, pp. 713–715, 1987.

[28] F. E. Gardiol, Wave propagation in a rectangular waveguide loaded with an H-plane dielectric slab, *IEEE Transactions on Microwave Theory and Techniques*, pp. 56–57, Jan. 1969.

[29] I. P. Polichronakis and S. S. Kouris, Computation of the dispersion characteristics of a shielded suspended substrate microstrip line, *IEEE Transactions on Microwave Theory and Techniques*, Vol. 40, No. 3, pp. 581–584, Aug. 1992.

[30] F. E. Gardiol, Careful MIC design prevents waveguide mode, *Microwaves Journal*, pp. 188–191, May 1977.

[31] D. Rubin and A. R. Hislop, Millimeter-wave coupled line filters: design techniques for suspended substrate and microstrip, *Microwave Journal*, Vol. 23, No. 10, pp. 67–78, Oct. 1980.

[32] C. I. Mobbs and J. D. Rhodes, A generalized Chebyshev suspended substrate stripline bandpass filter, *IEEE Transactions on Microwave Theory and Techniques*, Vol. 31, No. 5, pp. 397–402, May 1983.

[33] S. A. Alseyab, A novel class of generalized Chebyshev low-pass prototype for suspended substrate stripline filters, *IEEE Transactions on Microwave Theory and Techniques*, Vol. 30, No. 9, pp. 1341–1347, Sept. 1982.

[34] J. D. Rhodes, Suspended substrate filters and multiplexers, *16th European Microwave Conference*, Dublin, Ireland, pp. 8–18, Sept. 8–12, 1986.

[35] W. Schwab, W. Menzel, and F. Boegelsack, Multilayer suspended stripline and coplanar line filters, *IEEE Transactions on Microwave Theory and Techniques*, Vol. 42, No. 7, pp. 1403–1407, July 1994.

[36] W. Menzel and W. Schwab, Compact bandpass filters with improved stop-band characteristics using planar multilayer structures, *IEEE International Microwave Symposium Digest*, Vol. 3, pp. 1207–1209, 1992.

[37] R. M. Dougherty, Mm-wave filter design with suspended stripline, *Microwave Journal*, pp. 75–84, July 1986.

[38] V. Kotchetygov and S. H. H. Lim, Suspended substrate stripline filter design with hybrid EM simulation tools, *Proceedings of the Asia Pacific Microwave Conference*, pp. 1191–1194, Dec. 2000.

[39] A. Hislop and D. Rubin, Suspended substrate Ka-band multiplexer, *Microwave Journal*, pp. 73–76, June 1981.

[40] M. Miyazaki, Y. Isota, N. Takeuchi, and O. Ishida, An asymmetrical suspended stripline directional coupler, *Electronics and Communications in Japan*, Pt. 2, Vol. 81, No. 12, pp. 58–63, Jan. 1999.

[41] J. E. Dean and J. D. Rhodes, Design of MIC broadband contiguous multiplexers, *Proceedings of the 9th European Microwave Conference*, pp. 407–411, 1979.

[42] A. Brown and G. M. Rebeiz, A varactor tuned RF filter, *IEEE Transactions on Microwave Theory and Techniques*, Vol. 48, No. 7, pp. 1157–1160, July 2000.

[43] Y. C. Chiang, C. K. C. Tzuang, and S. Su, Design of a three-dimensional gap-coupled suspended substrate stripline bandpass filter, *IEE Proceedings H*, Vol. 139, No. 4, pp. 376–384, Aug. 1992.

[44] D. F. Argollo, H. Abdalla, and A. J. M. Soares, Method of lines applied to broadside suspended stripline coupler design, *IEEE Transactions on Magnetics*, Vol. 31, No. 3, pp. 1634–1636, May 1995.

[45] W. Menzel and F. Bogelsack, Folded stubs for compact suspended stripline circuits, *IEEE MTT-S Digest*, pp. 593–596, 1993.

[46] W. Menzel, A novel miniature suspended stripline filter, *Proceedings of the 33rd European Microwave Conference*, Munich, Germany, pp. 1047–1050, 2003.

[47] S. Uysal and L. Lee, Ku-band double-sided suspended substrate microstrip coupled-line bandpass filter, *Electronics Letters*, Vol. 35, No. 13, pp. 1088–1090, June 1999.

[48] R. Tahim, G. Hayashibara, and K. Chang, Design and performance of W-band broad-band integrated circuit mixers, *IEEE Transactions on Microwave Theory and Techniques*, Vol. 31, No. 3, pp. 277–283, Mar. 1983.

[49] E. Yamashita, K. Atsuki, B. Wang, and K. R. Li, Effects of side-wall grooves on transmission characteristics of suspended striplines, *IEEE Transactions on Microwave Theory and Techniques*, Vol. 33, No. 12, pp. 1323–1328, Dec. 1985.

[50] J. C. Goswani and M. Sachidananda, Cylindrical cavity-backed suspended stripline antenna: theory and experiment, *IEEE Transactions on Antennas and Propagation*, Vol. 41, No. 8, pp. 1155–1160, Aug. 1993.

[51] X. Jinxiong, Suspended stripline and Ka band integrated mixer, *Asia Pacific Conference on Environmental Electromagnetics*, Shanghai, China, May 2000.

[52] S. K. Koul, R. Khanna, and B. Bhat, Characteristics of single and coupled rectangular resonator in suspended substrate stripline, *Electronics Letters*, Vol. 22, No. 7, pp. 376–378, Mar. 1986.

[53] C. Cho and K. C. Gupta, Design methodology for multilayer coupled line filters, *IEEE International Microwave Symposium Digest*, Vol. 2, pp. 785–788, 1997.

CHAPTER 4

MINIATURIZATION OF PLANAR RESONATORS USING FRACTAL ITERATIONS

4.1 Introduction 79
4.2 Miniaturization of planar resonators 81
4.3 Fractal iteration applied to planar resonators 82
4.4 Minkowski resonators 84
  4.4.1 Minkowski's fractal iteration 84
  4.4.2 Minkowski square first-iteration resonator 85
  4.4.3 Minkowski rectangular first-iteration resonator 86
  4.4.4 Minkowski second-iteration resonators 88
  4.4.5 Sensitivity of minkowski resonators 90
4.5 Hibert resonators 92
  4.5.1 Hilbert's curve 92
  4.5.2 Hilbert half-wavelength resonator 93
  4.5.3 Unloaded quality factor of hilbert resonators 94
  4.5.4 Sensitivity of hilbert resonators 96
  References 98

4.1 INTRODUCTION

In the late nineteenth century, several mathematicians were interested in difficult mathematical challenges such as continuous but nondifferentiable curves with infinite length and finite occupied area. These curves are defined as the

Design and Realizations of Miniaturized Fractal RF and Microwave Filters,
By Pierre Jarry and Jacques Beneat

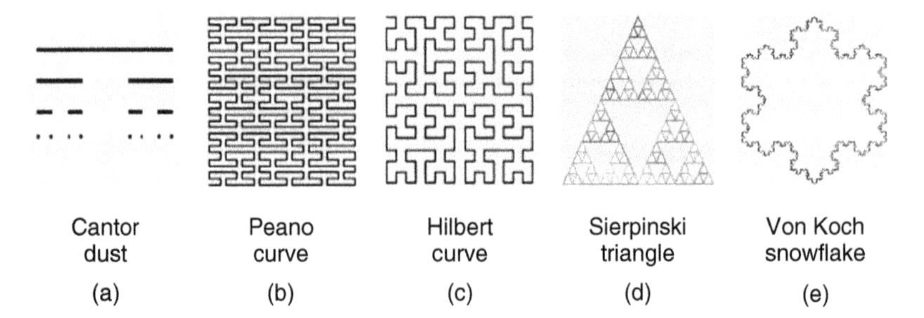

Figure 4.1 Examples of fractal construction.

infinite limit of an iterative construction process. Cantor probably was the first to describe fractal iteration in the 1870s [Figure 4.1(a)]. In 1890, Peano published another form of fractal iterations [Figure 4.1(b)]. Hilbert's eponymous fractal [Figure 4.1(c)] appeared in 1891, Sierpinsky's carpet [Figure 4.1(d)] in 1915, and Von Koch's snowflake (a continuous curve with no tangent, obtained by an elementary geometrical construction) in 1904 [Figure 4.1(e)]. However, the term *fractal* (from the Latin adjective *fractus*, which means "irregular" or "broken") was not introduced until the 1970s in a book called *Fractals* [2] by the mathematician Benoit Mandelbrot. Today, fractals are used in many areas and their applications are diverse. Due to their self-similarity and space-filing properties, it is possible to model complex shapes present in nature that were impossible to describe with classical Euclidean geometry.

Fractals are used in many areas:

- Statistical analysis, for predicting extreme events
- Astronomy, to understand galaxy formation
- Chemistry, for homogeneous mixtures
- Geophysical analysis, to predict earthquakes and floods
- Acoustics, for the construction of soundproof walls
- Physiology, for an understanding of the lungs' morphology
- Computer science, for numerical analysis and image compression

In the area of electromagnetic field propagation, fractals have been used to design compact and multiband antennas. In the past few years we have witnessed an increasing interest in fractals in the design of miniaturized filters. The following references provide a better understanding of fractal geometry as well as its application in the electromagnetic propagation: general principles, fractals, and electromagnetic propagation [1–9], fractal antennas [10–14], fractal resonators [15–18], periodic fractal structures [19,20], and fractal filters [21–24]. In the remainder of this chapter, we examine a few fractal iterations to determine their suitability as planar resonators for radio-frequency (RF) and microwave filters.

4.2 MINIATURIZATION OF PLANAR RESONATORS

Research on miniaturizing planar resonators has intensified over the past few years to answer the constant need for RF and microwave filters of smaller size and cost. Figure 4.2 shows a first example of miniaturization of a $\lambda/2$ line resonator at 7.3 GHz. By folding the line, one can achieve a reduction in the form factor while maintaining the resonator's length approximately. The resonator on the left in Figure 4.2 will be referred to a line resonator, the resonator in the middle as a square resonator, and the resonator on the right as a meandered line resonator. The dimensions of the square and meandered line resonators are modified so that all resonators have the same resonant frequency of 7.3 GHz. The line adjustments are to compensate for the various couplings introduced by the meanders during the miniaturization. In the examples that follow, the resonators are made using suspended substrate technology. The enclosure is a cavity with an upper height of 1 mm and a lower height of 1 mm. The substrate used is Duroid 5880, which has a relative permittivity of 2.2 (loss tangent 0.0009) and a thickness of 0.254 mm. The conducting strips used in the resonators are made of copper.

Figure 4.3 shows simulated transmission characteristics of the three resonators shown in Figure 4.2. The first spurious response of the line resonator occurs at twice the resonant frequency, while the first spurious response of the square and meandered line resonators occurs at about three times the resonant frequency. The spurious response at twice the resonant frequency has been suppressed in the case of the square and meandered line resonators. This is due to even- and odd-mode velocities being equalized. It is also seen that these two resonators can produce a transmission zero. This is due to the internal couplings within these resonators. On the other hand, the line resonator has the highest unloaded quality factor at $Q_0 = 308$, while $Q_0 = 280$ for the square resonator and $Q_0 = 259$ for the meandered line resonator. The degradation in the unloaded quality factor is due to the increase in current density in these resonators.

In the following figures, the sensitivity of these resonators to the geometrical variations in suspended substrate technology is examined. Figure 4.4(a) shows how variations in upper cavity height can change the resonant frequency of the three resonators. These simulations show that the line resonator and the square resonator exhibit similar behavior. Their resonant frequency increases when the upper height is lowered. The meandered line resonator shows an opposite effect.

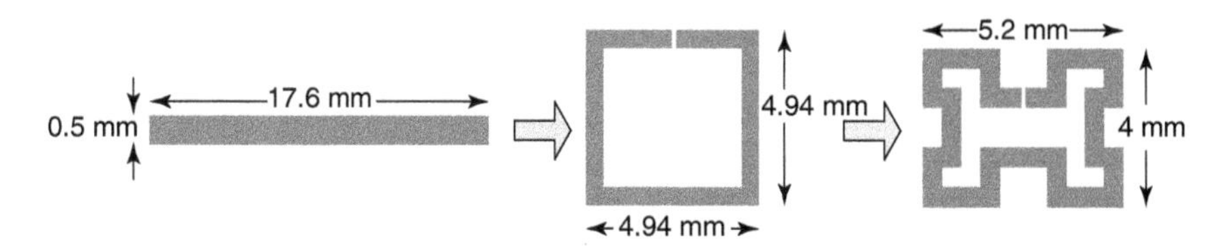

Figure 4.2 Miniaturization of a half-wavelength line resonator.

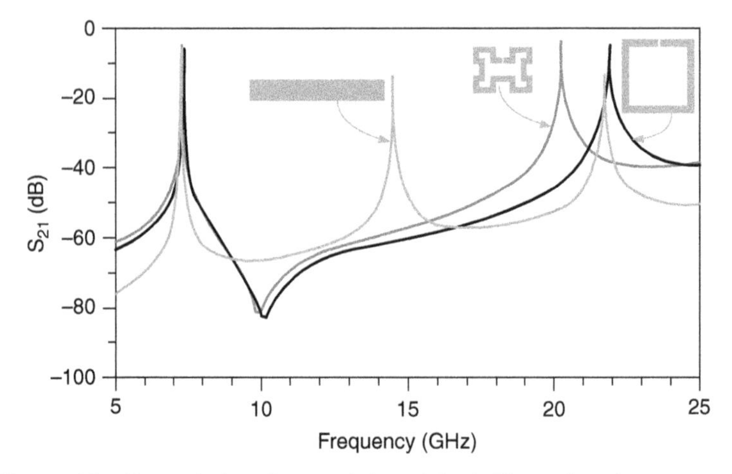

Figure 4.3 Transmission characteristics of the half-wavelength resonators.

Its resonant frequency increases as the upper height increases. The range of variations over the interval is slightly smaller for the meandered line resonator. Figure 4.4(b) shows how variations in the lower cavity height can change the resonant frequency of the three resonators. All three resonators exhibit the same behavior. The resonant frequency increases as the lower cavity height increases. The meandered line resonator has somewhat larger variations in the interval considered.

Figure 4.4(c) shows how variations in substrate thickness can change the resonant frequency of the three resonators. All three resonators exhibit closely the same behavior. The resonant frequency decreases as the substrate thickness increases.

In conclusion, Figure 4.4 shows that miniaturizing the resonator does result in resonators that are more sensitive to suspended substrate fabrication tolerances such as cavity heights and substrate thickness.

4.3 FRACTAL ITERATION APPLIED TO PLANAR RESONATORS

In Section 4.2 the miniaturized resonators had rather simple shapes. In this section the behavior of true fractal resonators is examined. For example, Figure 4.5 shows the reflection characteristics of several resonators obtained using fractal iterations. The additional miniaturization step to make them all have the same resonant frequency was not applied. Instead, the resonators in Figure 4.5 all have the same overall size. A fractal is obtained by successive applications of a modification to a shape. For example, in the top right case of Figure 4.5, the first motif is a filled square (rightmost motif). Resonators starting with such a surface are referred to as *patch resonators*. After the first fractal iteration, the second motif is the original square but with a small square removed from each

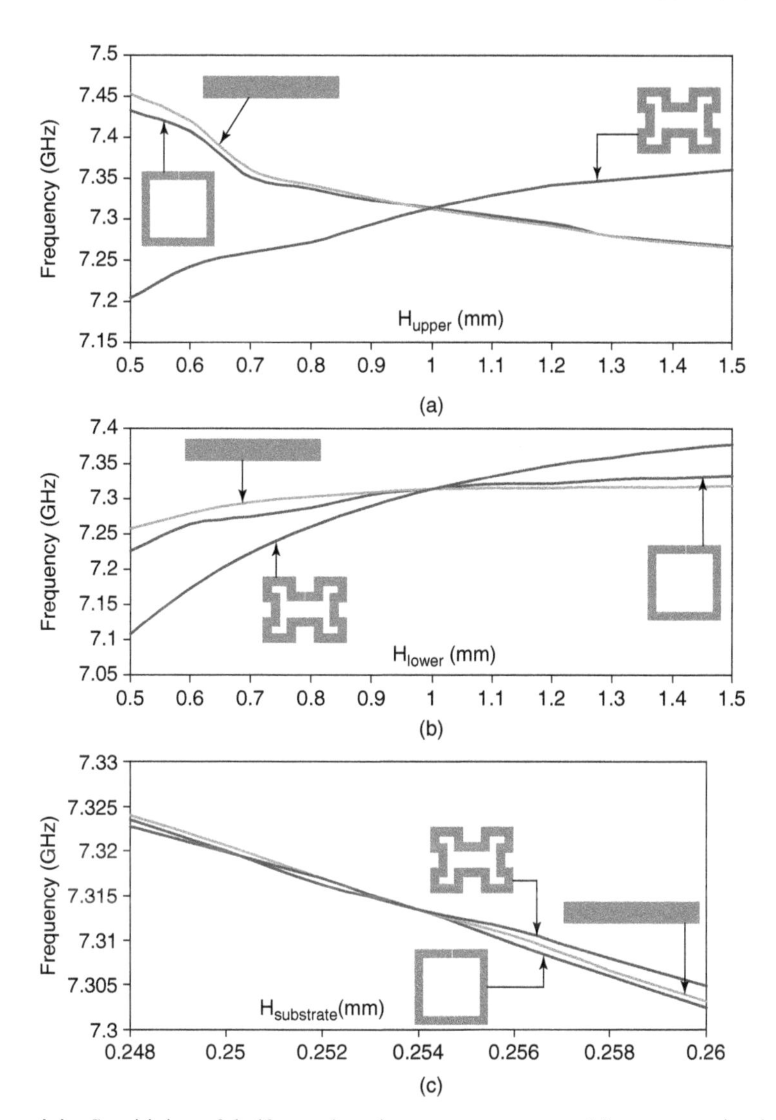

Figure 4.4 Sensitivity of half-wavelength resonators versus (a) upper cavity height, (b) lower cavity height, and (c) substrate thickness.

side of the square. One can check that the third motif is the result of applying this modification to each square present in the second motif.

The simulations of Figure 4.5 show clearly that when the fractal iteration order increases, the resonant frequency of the resonator decreases. This is a very important result since for a given resonant frequency, the higher the fractal iteration number, the smaller the size of the resonator. This means that at least for the fractal iterations shown in Figure 4.5, the use of fractal resonators compared to other types of resonators can lead to smaller filters.

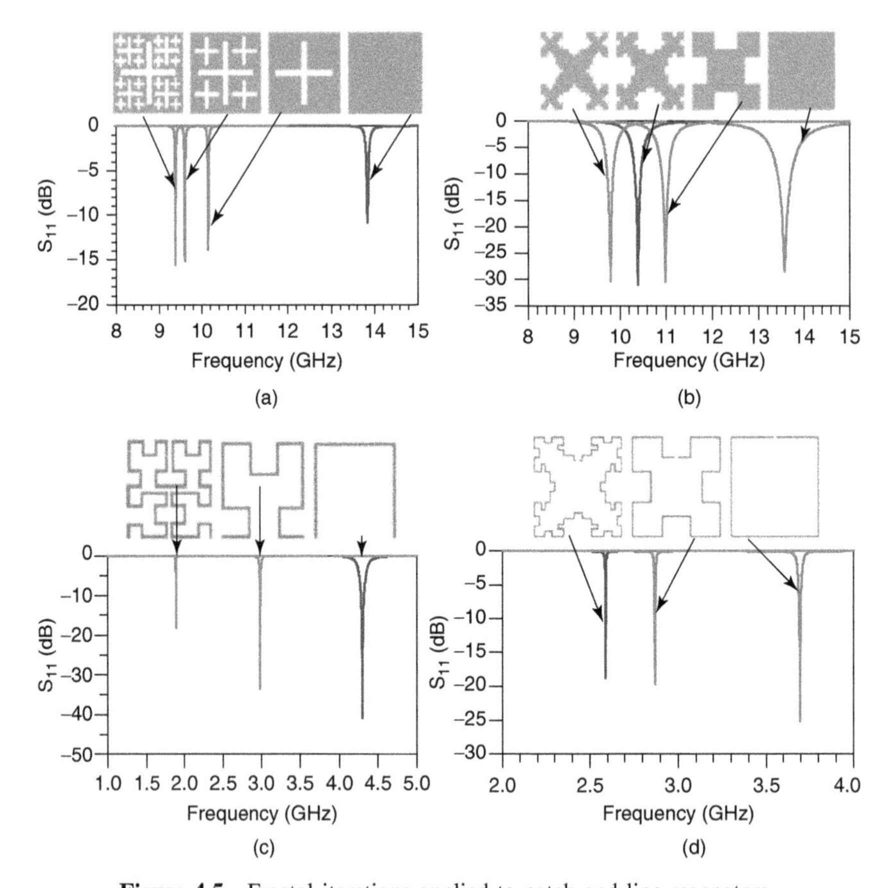

Figure 4.5 Fractal iterations applied to patch and line resonators.

Figure 4.5 shows also another very important result. The decrease in resonant frequency is most significant during the first two fractal iterations. For the third iteration and above, few changes are observed. This is important since it means that after two fractal iterations, the gain in miniaturization is minimal compared to the complexity of fabricating the resonator.

In the remainder of this chapter we show the suitability of two types of fractal resonators for miniaturized RF and microwave filters. The first type is based on Minkowski's fractal iterations, and the second type is based on Hilbert's fractal iterations.

4.4 MINKOWSKI RESONATORS

4.4.1 Minkowski's Fractal Iteration

Minkowski's fractal iteration technique is illustrated in Figure 4.6. There are two types of iterations to consider. The square iteration technique is based on

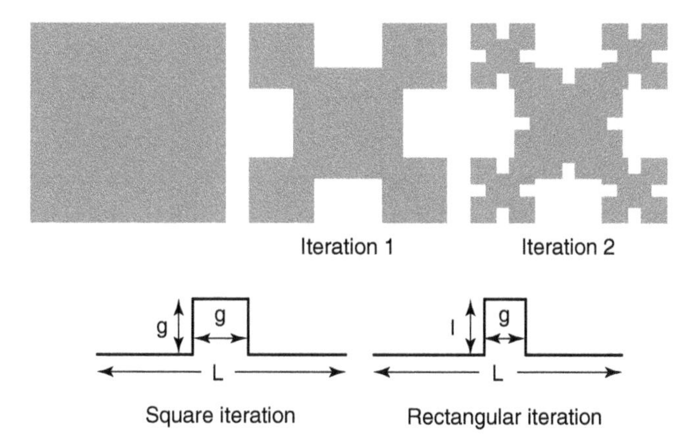

Figure 4.6 Minkowski's iteration applied to a square dual-mode resonator.

removing small squares (g), while the rectangular iteration technique is based on removing small rectangles (g and l). Both techniques start from a filled square with side length L. For the square iteration, after one iteration the motif corresponds to the original square (L), where a small square (g) was removed from each side. After two iterations, the motif is the result of removing a small square (g) from each side of every square in the motif of iteration 1. Note that the absolute value of g decreases after each iteration. What is preserved is the overall space (L). The motifs are referred to as *Minkowski's islands*. After an infinite number of iterations, there would be nothing left [it has consumed the entire space of the square (L)].

For the rectangular iteration, one removes small rectangles (g and l) from each side of the squares.

4.4.2 Minkowski Square First-Iteration Resonator

First we show how the size of the small square g can change the resonant frequency of the resonator. Figure 4.7 shows that the length of the original square L must be reduced as the value of g is increased, to keep the resonant frequency of the first-iteration resonator at 14 GHz. This is an important result since it shows that the amount of miniaturization could be controlled through the value of g. Next, we show how the size of the small square g can change the transmission characteristics of the resonator. Figure 4.8 shows the transmission characteristics of the initial square resonator and two first-iteration resonators with different g values. All resonators are designed to have the same resonant frequency of 14 GHz. Therefore, the first-iteration resonators have been miniaturized.

The first spurious response of the square resonator occurs at twice the resonant frequency, as expected. In addition to the size reduction, Minkowski's first-iteration resonators have a first spurious response that occurs clearly beyond that of the square resonator. It is also interesting to note that the first spurious

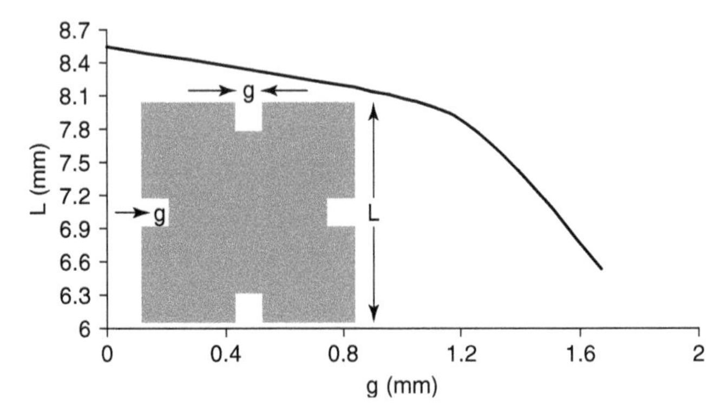

Figure 4.7 Relationship between L and g for a resonance at 14 GHz.

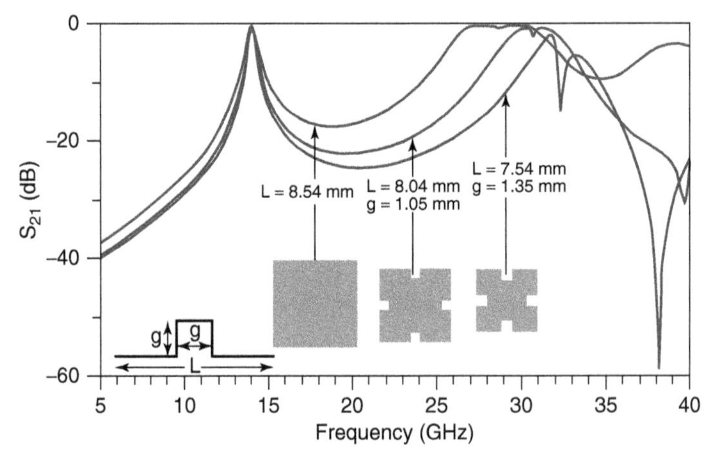

Figure 4.8 Transmission characteristics for square and first-iteration square Minkowski resonators.

response of the smallest Minkowski resonator occurs at the highest frequency, so that size and spurious response improvements go together.

4.4.3 Minkowski Rectangular First-Iteration Resonator

In this case a small rectangle of width g and height l is used instead of the small square g in the fractal iterations. In our examples g will have two possible values: $g = 1.05$ mm and $g = 1.35$ mm. To keep the resonant frequency at 14 GHz, both miniaturization (changing L) and adjusting l are used. The transmission characteristics for a few first-iteration rectangular Minkowski resonators are shown in Figure 4.9. We observe that in the case of rectangular Minkowski resonators, the first spurious response also occurs beyond twice the resonant

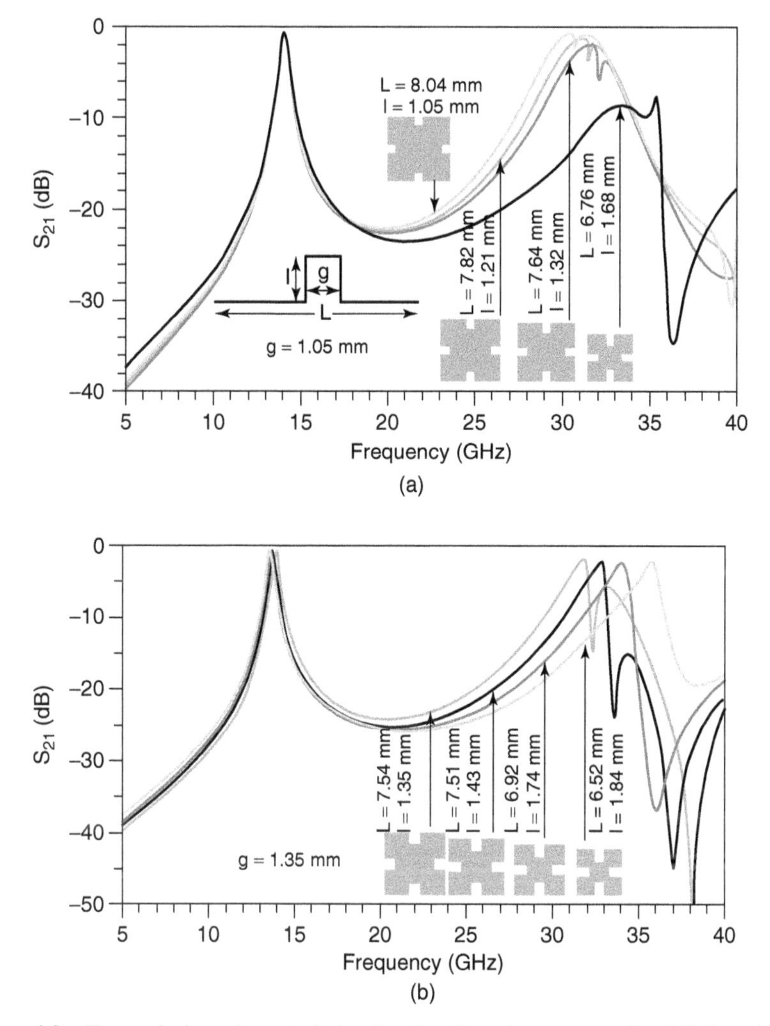

Figure 4.9 Transmission characteristics for first-iteration rectangular Minkowski resonators: (a) $g = 1.05$ mm; (b) $g = 1.35$ mm.

frequency. Figure 4.9 also confirms that the first spurious response improves as the size of the resonator decreases.

To compare square and rectangular Minkowski resonators, Figure 4.10 shows the transmission characteristics of first-iteration square and rectangular resonators. The two resonators are based on the same initial space ($L = 6.76$ mm). The square iteration used $g = 1.56$ mm. The rectangular iteration used $g = 1.05$ mm and $l = 1.68$ mm. The simulations show that the rectangular resonator has a spurious response that is attenuated by about 8 dB compared to the square resonator. However, computations of the unloaded quality factor show no significant difference ($Q_u = 424$ for the rectangular, $Q_u = 438$ for the square). The computation

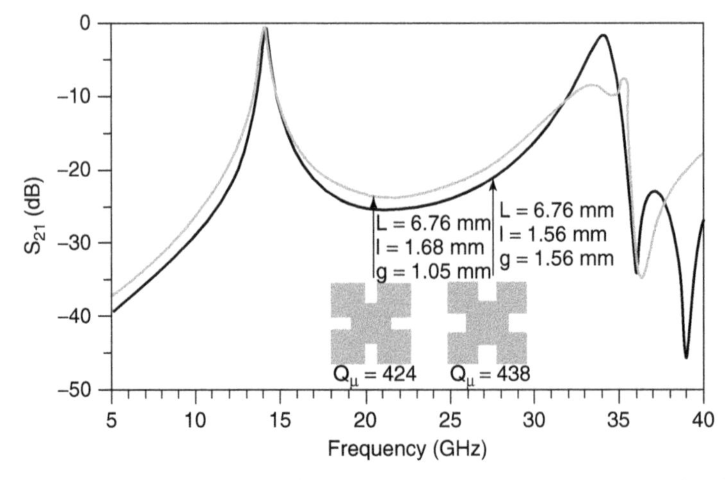

Figure 4.10 Transmission characteristics of a square versus a rectangular first-iteration Minkowski resonator.

technique used to define the value of the unloaded quality factor is described in Appendix 2.

4.4.4 Minkowski Second-Iteration Resonators

For a higher level of miniaturization we examine the characteristics of second-iteration Minkowski resonators. Figure 4.11 shows the transmission characteristic of the initial square resonator and first- and second-iteration Minkowski resonators. Miniaturization was used, so that all three resonators have

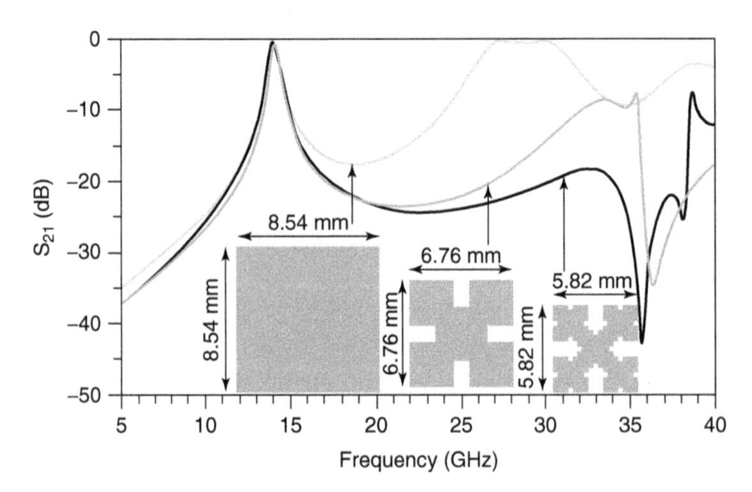

Figure 4.11 Transmission characteristics of square and first- and second-iteration Minkowski resonators.

a resonant frequency of 14 GHz. Besides the reduction in size due to increased fractal iteration, Figure 4.11 shows that the transmission characteristics of the second-iteration resonator are better than that of the first-iteration resonator. For example, the first spurious response of the second-iteration resonator is attenuated compared to that of the first-iteration resonator and occurs far beyond twice the resonant frequency, as is the case for the square resonator. In addition, the second-iteration resonator produces a clearer transmission zero.

Table 4.1 gives the unloaded quality factor of the three resonators of Figure 4.11. The unloaded quality factor is seen to be better for the second-iteration resonator. This is remarkable since one would expect the unloaded quality factor to get worst when the resonator is miniaturized. When the size of the resonator decreases, the radiation loss decreases but the conduction loss increases, since the current density increases, thus leading to a lower unloaded quality factor.

One possible explanation for the results of Table 4.1 is that the second-iteration resonator has a longer perimeter than the others. The energy stored in a resonator is concentrated primarily on its boundaries. This could explain why the second-iteration resonator has a higher unloaded quality factor. Figure 4.12 shows the current distribution for each of the three resonators at the resonant frequency. One can see that the current is distributed primarily on the boundaries.

Figure 4.13 shows the transmission characteristics of second-iteration square Minkowski resonators. The resonators are designed to have the same resonant frequency of 14 GHz. Looking at the resonators in Figure 4.13, one can see that the value of g used in the first iteration is quite different from the value of g used in the second iteration. One also observes that resonators with larger values of g result in smaller resonators. This is consistent with the results of Figure 4.7.

Figure 4.13 shows that the first spurious response occurs far beyond twice the resonant frequency. It also shows that second-iteration resonators are capable of

TABLE 4.1 Unloaded Quality Factor of the Resonators of Figure 4.11

Resonator	Size (mm^2)	Perimeter (mm)	Q_0
	8.54 × 8.54	34.16	353.84
	6.76 × 6.76	40.48	424.3
	5.82 × 5.82	48.56	425.8

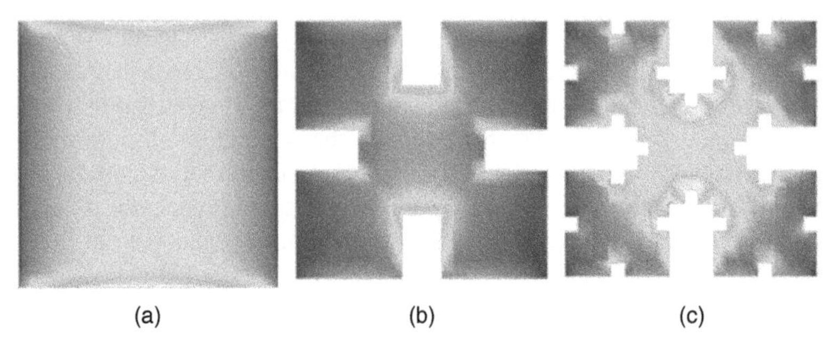

Figure 4.12 Current distribution in the three resonators of Figure 4.11.

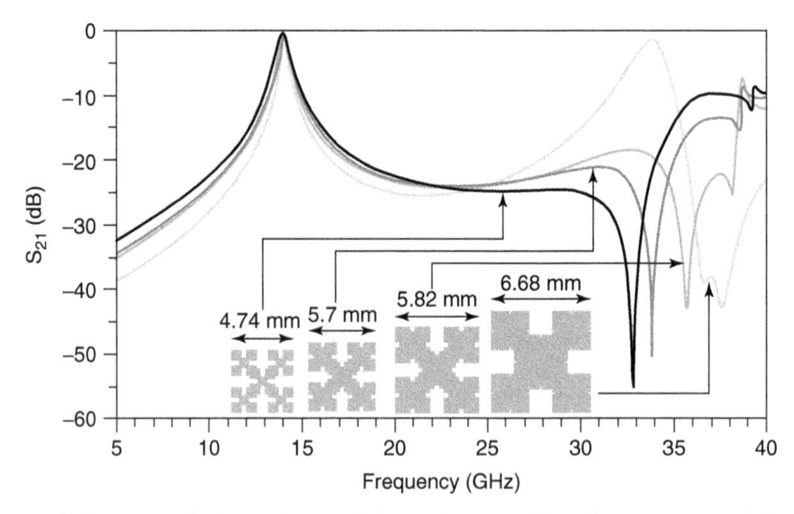

Figure 4.13 Transmission characteristics of second-iteration square Minkowski resonators.

producing a transmission zero. Table 4.2 provides the unloaded quality factors of these resonators as well as additional details on the dimensions and perimeters. As expected, the unloaded quality factor decreases when the size of the resonator decreases, except for resonator 1. However, one should note that the complexity of the pattern, or the amount of surface removed in resonator 1, is more negligible than in the other resonators.

4.4.5 Sensitivity of Minkowski Resonators

Fabrication tolerances are a problem in suspended substrate technology. Therefore, it is important to check how parameter variations could affect the performance of fractal resonators. Figure 4.14 shows how variations in the upper

TABLE 4.2 Unloaded Quality Factor of the Minkowski Resonators of Figure 4.13

Resonator	Dimensions (mm^2)	Perimeter (mm)	Unloaded Quality Factor Q_u
1	6.68×6.68	47.04	398
2	5.82×5.82	48.56	425.8
3	5.7×5.7	51.6	403.4
4	4.74×4.74	54.16	344.1

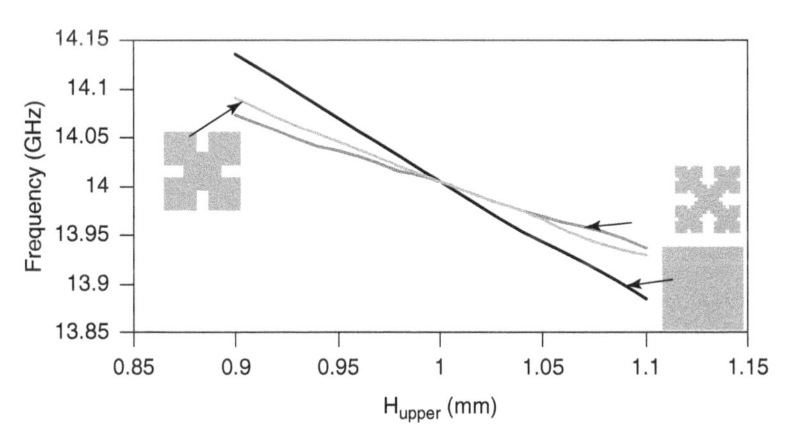

Figure 4.14 Sensitivity to upper cavity height variations.

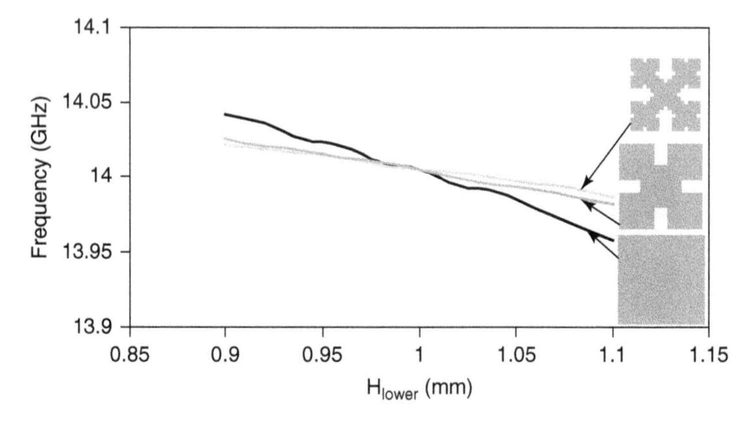

Figure 4.15 Sensitivity to lower cavity height variations.

cavity height specified can change the resonant frequency of a square and first- and second-iteration Minkowski resonators. All resonators are designed to have the same resonant frequency of 14 GHz when upper and lower cavity heights are 1 mm. It is seen that for a 0.2-mm variation in upper cavity height, the resonant frequency shift is about 1.78% for the square resonator, 1.15% for the first-iteration resonator, and 0.96% for the second-iteration resonator. This suggests that higher-iteration fractal resonators are less sensitive to upper cavity height variations.

Figure 4.15 shows how variations in the specified lower cavity height can change the resonant frequency of a square resonator and first- and second-iteration Minkowski resonators. For a 0.2-mm variation in lower cavity height, the resonant frequency shift is about 0.59% for the square resonator, 0.31% for the first-iteration resonator, and 0.29% for the second-iteration resonator. This suggests that higher-iteration fractal resonators are less sensitive to lower cavity height variations.

Figure 4.16 shows how variations in the specified substrate thickness can change the resonant frequency of a square resonator and first- and second-iteration Minkowski resonators. In this case, all resonators exhibit the same behavior when the substrate thickness is changed. Overall, these results show that Minkowski fractal resonators are great candidates for suspended substrate technology given their weak sensitivity to fabrication tolerances.

4.5 HIBERT RESONATORS

4.5.1 Hilbert's Curve

The Hilbert resonators used in this book are based on Hilbert's curve. It is obtained using the fractal iterations shown in Figure 4.17. A U shape (iteration 0) in a motif at a given iteration gets replaced by the pattern shown in iteration 1.

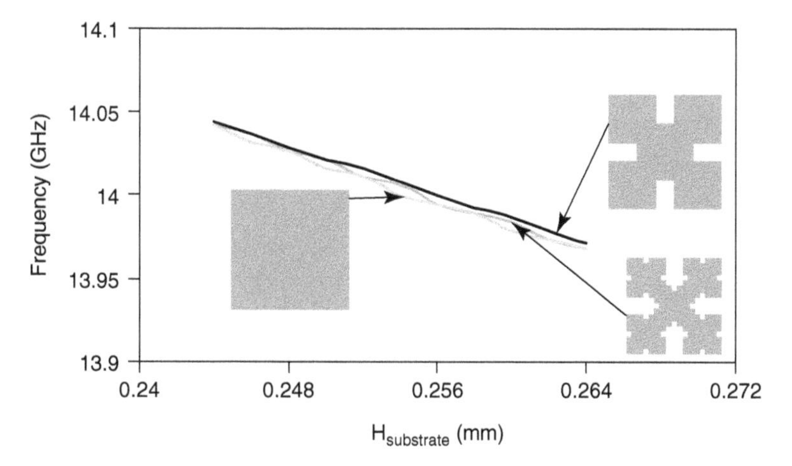

Figure 4.16 Sensitivity to substrate thickness.

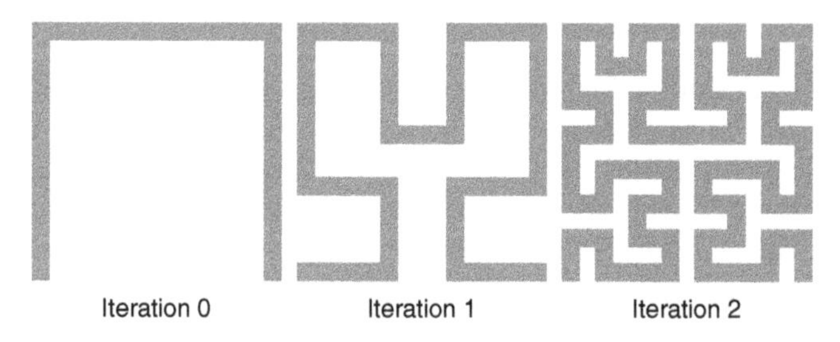

Figure 4.17 Hilbert's curve fractal iteration.

Repeating the fractal iterations to infinity, the Hilbert curve will fill in entirely the square surface that delimitates it. Therefore, in theory it is possible to fit a line of infinite length into a finite surface.

4.5.2 Hilbert Half-Wavelength Resonator

We are interested in examining the behavior of a $\lambda / 2$ line resonator to which we apply Hilbert's fractal iterations (Figure 4.18). The resulting resonators are

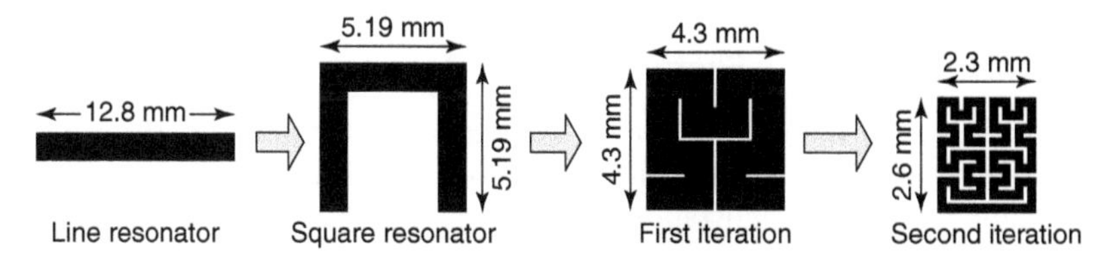

Figure 4.18 Hilbert iteration applied to a $\lambda/2$ resonator.

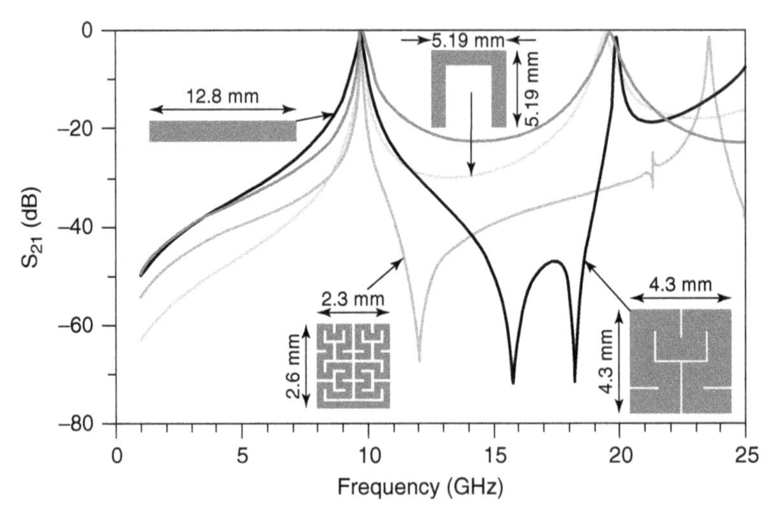

Figure 4.19 Transmission characteristics of Hilbert resonators at 9.78 GHz.

miniaturized so that all resonators have the same resonant frequency of 9.78 GHz. The transmission characteristics for line, square, and first- and second-iteration Hilbert resonators designed for a resonant frequency of 9.78 GHz are given in Figure 4.19. The first spurious response occurs at twice the resonant frequency for line and square resonators. It occurs slightly above twice the resonant frequency for a first-iteration Hilbert resonator and clearly above twice the resonant frequency for a second-iteration Hilbert resonator. The first- and second-iteration Hilbert resonators also produce transmission zeros. The zero produced by the second-iteration resonator is, however, closer to the resonant frequency (passband). One should note that the miniaturization applied on the second-iteration Hilbert resonator has reduced the line width. This has the effect of increasing the ohmic losses.

The transmission characteristics for square and first- and second-iteration Hilbert resonators designed for a resonant frequency of 6.9 GHz are given in Figure 4.20. In this case, all resonators have the same line width of 0.2 mm. In this case again, the first spurious response is higher than twice the resonant frequency for the first- and second-iteration Hilbert resonators. It also confirms that the transmission zero is closer to the resonant frequency for the higher-iteration Hilbert resonator, resulting in improved selectivity.

4.5.3 Unloaded Quality Factor of Hilbert Resonators

In this section we discuss the unloaded quality factor for the Hilbert resonators and compare it to other shaped planer resonators. The losses of the structure are both metallic (cavities, metallization) and dielectric (substrate). The conductivity

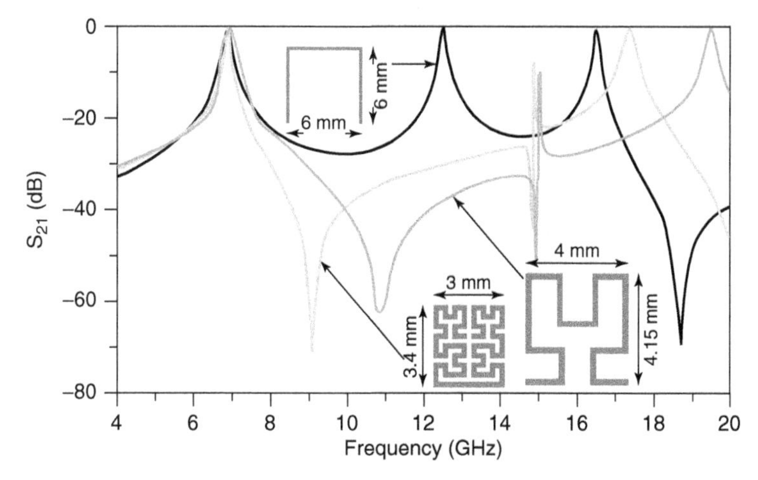

Figure 4.20 Transmission characteristics of Hilbert resonators at 6.9 GHz.

TABLE 4.3 Unloaded Quality Factor of the 6.9-GHz Hilbert Resonators

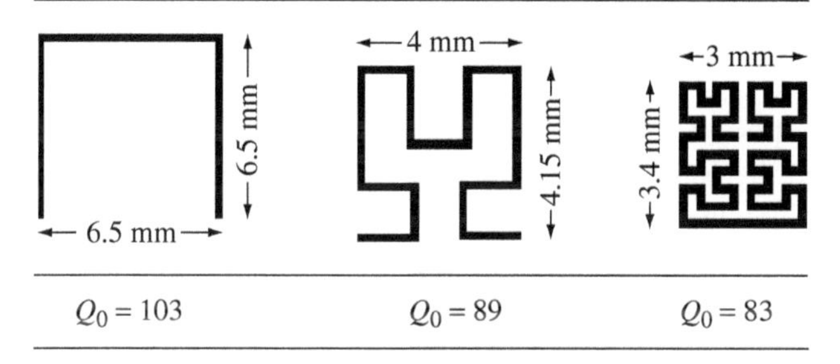

$Q_0 = 103$	$Q_0 = 89$	$Q_0 = 83$

of copper ($\sigma = 5.8 \times 10^7$ S/m) characterizes the metallic losses. The dielectric losses are characterized by the substrate's loss tangent ($\tan\sigma = 9 \times 10^{-4}$). The substrate has a dielectric constant of 2.2 and a thickness of 254 μm. The upper and lower cavity heights are both equal to 1 mm. Table 4.3 reports the unloaded quality factors for the square and first- and second-iteration Hilbert resonators. All resonators are designed for a resonant frequency of 6.9 GHz.

It is seen that the unloaded quality factor decreases as the size of the resonator decreases. To retain the same resonant frequency, the first- and second-iteration resonators must be miniaturized. This has the effect of increasing the metallic losses and therefore degrading the quality factor. In Table 4.4 we compare the unloaded quality factors and resonant frequencies of the second-iteration Hilbert resonator with two other shaped resonators. All resonators are designed to use the same space and line length. The first resonator resonates at 5.48 GHz with

TABLE 4.4 Unloaded Quality Factors of Three Resonators Occupying the Same Space

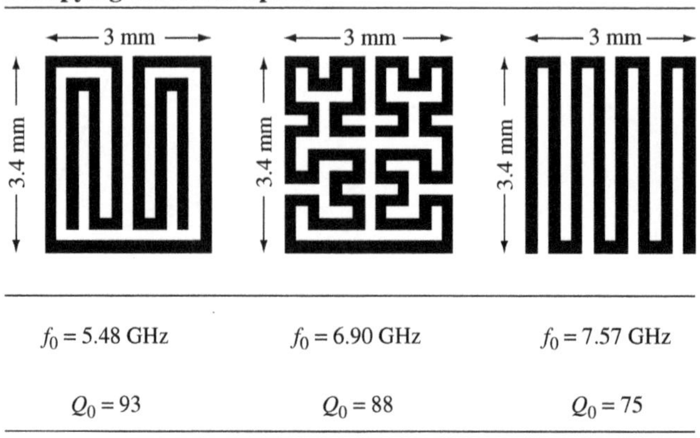

$f_0 = 5.48$ GHz	$f_0 = 6.90$ GHz	$f_0 = 7.57$ GHz
$Q_0 = 93$	$Q_0 = 88$	$Q_0 = 75$

an unloaded quality factor $Q_0 = 93$, the second-iteration Hilbert resonator resonates at 6.9 GHz with $Q_0 = 88$, and the third resonator resonates at 7.57 GHz with $Q_0 = 75$. The difference in resonant frequency is probably the result of "electromagnetic shortcuts" produced by couplings within the meanders of the resonators. These shortcuts would be responsible for shorting the effective length of the resonator, hence increasing its resonant frequency.

The rather low unloaded quality factors are a result of using copper as the conductor and the rather large substrate thickness. To improve the electrical performance such as the quality factor, one could use a better-conducting material such as gold, a thinner substrate, or a substrate with a smaller loss tangent. In addition, fractal resonators are suitable for other technologies that allow a significant enhancement of the unloaded quality factor, such as micromachining and superconductivity. The small size of fractal resonators can reduce the cost of the expensive substrate and satisfy the small substrate-size limitations used in these technologies.

4.5.4 Sensitivity of Hilbert Resonators

The sensitivity of Hilbert resonators regarding possible variations of the geometrical parameters of suspended substrate technology is examined next. Figure 4.21 shows the resonant frequency variations as a result of changes in upper and lower cavity height and in substrate thickness. The results are provided for line, square, and first-iteration Hilbert resonators. The three resonators are designed to have a resonant frequency $f_0 = 9.78$ GHz when the upper and lower cavity heights are equal to 1 mm and the substrate thickness is equal to 0.254 mm. One can see that the first-iteration Hilbert resonator is less sensitive to upper cavity height variations, is slightly more sensitive to lower cavity-height variations, and

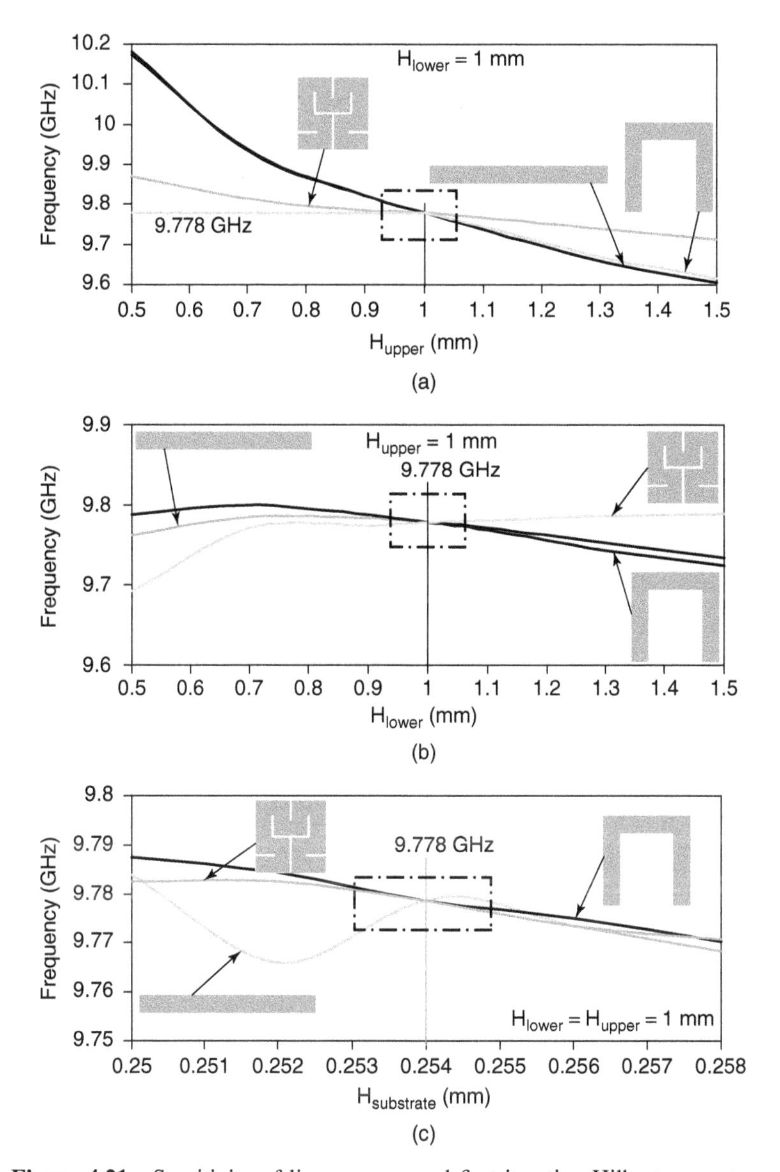

Figure 4.21 Sensitivity of line, square, and first-iteration Hilbert resonators.

has about the same sensitivity to substrate thickness variations as the other two resonators.

Figure 4.22 shows variations in resonant frequency when the upper and lower cavity heights as well as the substrate thickness are changed for first- and second-iteration Hilbert resonators. One can see that the second-iteration Hilbert resonator seems to be less sensitive than the first-iteration Hilbert resonator to all three types of variations (upper and lower cavity height and substrate thickness).

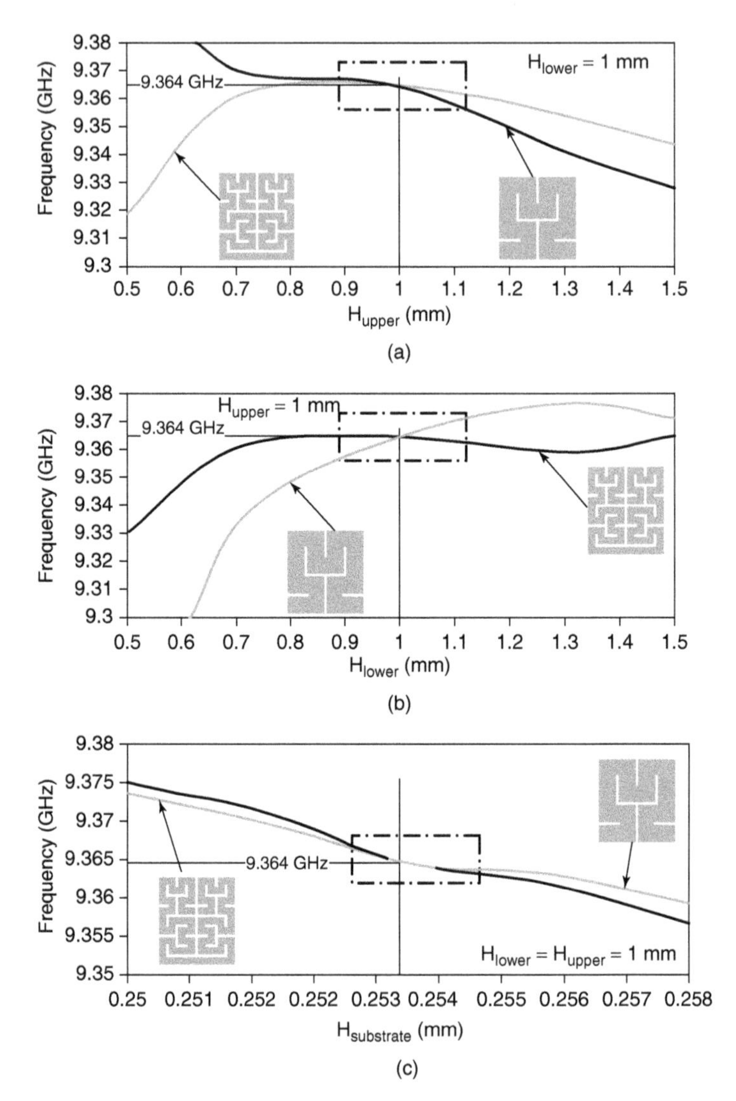

Figure 4.22 Sensitivity of first- and second-iteration Hilbert resonators.

This tends to indicate that the higher the iteration number, the less sensitive the resonator to suspended substrate geometrical variations.

REFERENCES

[1] B. Mandelbrot, *The Fractal Geometry of Nature*, W.H. Freeman, New York, 1982.

[2] B. Mandelbrot, *Fractals: Form, Chance and Dimension*, W.H. Freeman, New York, 1977.

[3] Q. J. Zhang and K. C. Gupta, *Neural Networks for RF and Microwave Design*, Artech House, Norwood, MA, 2000.

[4] J. Levy-Veheh, E. Lutton, and C. Tricot, *Fractals in Engineering*, Springer-Verlag, New York, 1997.

[5] D. H. Werner and R. Mittra, *Frontiers in Electromagnetics*, IEEE Press Series on Microwave Technology and RF, Wiley–IEEE Press, New York, November 1999.

[6] M. W. Takeda, S. Kirihara, Y. Miyamoto, K. Sakuda, and K. Honda, Localization of electromagnetic waves in three-dimensional fractal cavities, *Physical Review Letters*, Vol. 92, No. 9, Mar. 2004.

[7] Y. Hobiki, K. Yakubo, and T. Nakayama, Spectral characteristics in resonators with fractal boundaries, *Physical Review Letters*, Vol. 54, No. 2 Aug. 1996.

[8] J. M. Rius, J. M. Gonzalez, J. Romeu, and A. Cardama, Conclusions and results of the FractalComs Project: exploring the limits of fractal electrodynamics for the future telecommunication technologies, http://www.tsc.upc.es/fractalcoms.

[9] Fractal miniaturization in RF and microwave networks, http://www.fractus.com.

[10] D. H. Werner, P. L. Werner, and A. J. Ferraro, Frequency independent features of self-similar antennas, *IEEE Antennas and Propagation Society*, Vol. 3, pp. 2050–2053, July 1996.

[11] C. Puente, J. Romeu, R. Pous, X. Garcia, and F. Benitez, Fractal multiband antenna based on the Sierpinski gasket, *Electronics Letters*, Vol. 32, pp. 1–2, Jan. 1996.

[12] C. Puente, J. Romeu, R. Pous, and A. Cardama, On the behavior of the Sierpinski multiband fractal antenna, *IEEE Transactions on Antennas and Propagation*, Vol. 46, No. 4, pp. 517–524, Apr. 1998.

[13] R. Breden and R. J. Langley, Printed fractal antennas, *IEEE National Conference on Antennas and Propagation*, Apr. 1999.

[14] J. Chang, S. Jung, and S. Lee, Triangular fractal antenna, *Microwave and Optical Technology Letters*, Vol. 27, No. 1, pp. 41–46, Oct. 2000.

[15] V. Crnojevic-Bengin and D. Budimir, Novel compact resonators with multiple 2-D Hilbert fractal curves, *Proceedings of the 35th European Microwave Conference*, Paris, pp. 205–207, 2005.

[16] V. Crnojevic-Bengin and D. Budimir, End-coupled microstrip resonator with multiple square Sierpinski fractal curves, *Proceedings of the Conference for Electronics, Telecommunications, Computers, Automation and Nuclear Engineering*, 2005.

[17] V. Crnojevic-Bengin and D. Budimir, Novel 3-D Hilbert microstrip resonators, *Microwave and Optical Technology Letters*, Vol. 46, No. 3, pp. 195–197, Aug. 2005.

[18] P. Pribetich, Y. Combet, G. Giraud, and P. Lepage, Quasifractal planar microstrip resonators for microwave circuits, *Microwave and Optical Technology Letters*, Vol. 21, No. 6, pp. 433–436, June 1999.

[19] F. Frezza, L. Pajewski, and G. Schettini, Fractal two-dimensional electromagnetic band-gap structures, *IEEE Transactions on Microwave Theory and Techniques,* Vol. 52, No. 1, pp. 220–223, Jan. 2004.

[20] K. Siakavara, Novel microwave microstrip filters using photonic bandgap ground plane with fractal periodic pattern, *Microwave and Optical Technology Letters*, Vol. 43, No. 4, pp. 273–276, Nov. 2004.

[21] M. Barra, C. Collado, J. Mateu, and J. M. O'Callaghan, Hilbert fractal curves for HTS miniaturized filters, *Proceedings of the International Microwave Symposium*, Fort Worth, TX, pp. 123–127, 2004.

[22] A. Prigiobbo, M. Barra, A. Cassinese, and M. Cirillo, Superconducting resonators for telecommunication application based on fractal layout, *Superconductor Science and Technology*, Vol. 17, pp. 427–431, 2004.

[23] P. Vincent, J. Culver, and S. Eason, Meandered line microstrip filter with suppression of harmonic passband response, *IEEE International Microwave Symposium Digest*, Vol. 3, No. 1, pp. 1905–1908, 2003.

[24] I. K. Kim, N. Kingsley, and M. Morton, Fractal-shape 40 GHz microstrip bandpass filter on hi-resisitivity Si for suppression of the 2nd harmonic, *Proceedings of the 35th European Microwave Conference*, Paris, pp. 825–828, 2005.

CHAPTER 5

DESIGN AND REALIZATIONS OF MEANDERED LINE FILTERS

5.1 Introduction 102

5.2 Third-order pseudo-elliptic filters with transmission zero on the right 102

5.2.1 Topology of the filter 102

5.2.2 Synthesis of the bandpass prototype 103

5.2.3 Defining the dimensions of the enclosure 103

5.2.4 Defining the gap distances 105

5.2.5 Further study of the filter response 107

5.2.6 Realization and measured performance 110

5.2.7 Sensitivity analysis 114

5.3 Third-order pseudo-elliptic filters with transmission zero on the left 117

5.3.1 Topology of the filter 117

5.3.2 Synthesis of the bandpass prototype 118

5.3.3 Defining the dimensions of the enclosure 118

5.3.4 Defining the gap distances 118

5.3.5 Realization and measured performance 122

5.3.6 Sensitivity analysis 125

References 127

Design and Realizations of Miniaturized Fractal RF and Microwave Filters,
By Pierre Jarry and Jacques Beneat

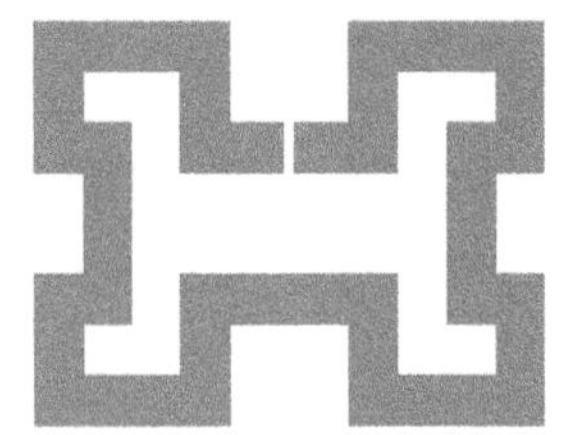

Figure 5.1 Half-wavelength meandered line resonator.

5.1 INTRODUCTION

In this chapter we provide the synthesis and design techniques used for filters using half-wavelength meandered line resonators. The resonator was introduced in Chapter 4 and is shown in Figure 5.1.

5.2 THIRD-ORDER PSEUDO-ELLIPTIC FILTERS WITH TRANSMISSION ZERO ON THE RIGHT

5.2.1 Topology of the Filter

The topology of a third-order pseudo-elliptic filter with a transmission zero on the right side of the passband is presented in Figure 5.2. The filter consists of three half-wavelength meandered line resonators. The transmission zero occurs on the right side of the passband, due to a magnetic coupling between resonators 1 and 3. Impedance matching for the 50-Ω input/output lines is achieved through an impedance transformer.

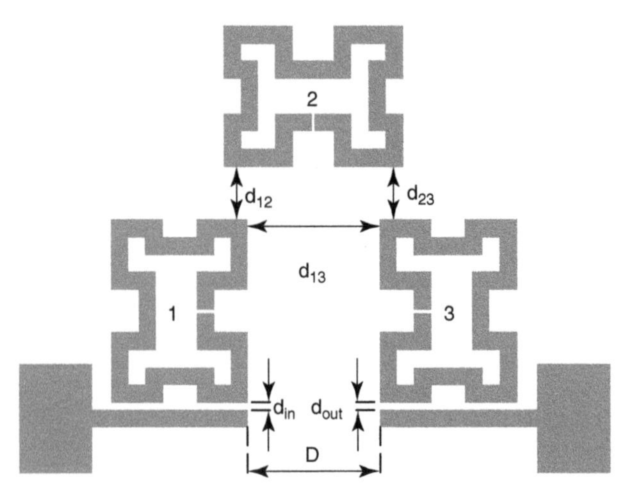

Figure 5.2 Topology of the third-order pseudo-elliptic filter with a transmission zero on the right.

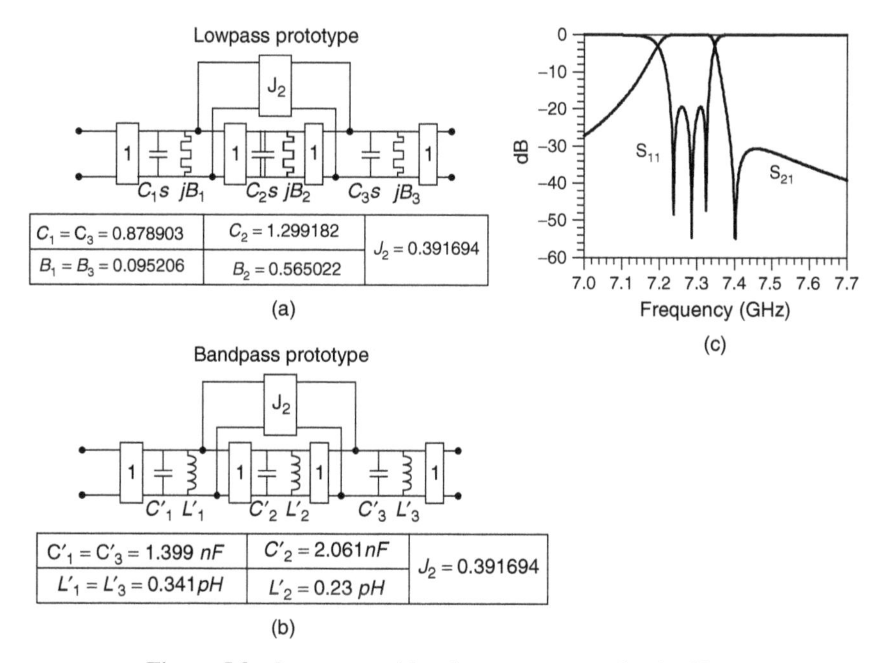

Figure 5.3 Lowpass and bandpass prototypes for the filter.

5.2.2 Synthesis of the Bandpass Prototype

The design of the filter starts with the synthesis of the bandpass pseudo-elliptic prototype circuit capable of producing the desired filtering characteristics. For example, we are interested in a third-order pseudo-elliptic filter centered at 7.28 GHz with a bandwidth of 100 MHz (1.37%), having a transmission zero on the right side of the passband at 7.4 GHz. Figure 5.3 shows the lowpass and bandpass third-order pseudo-elliptic prototypes and the bandpass transmission and reflection responses. The theoretical couplings needed for the design of the filter are $K_{12} = 0.0111$, $K_{13} = 0.0025$, and $K_{23} = 0.0111$. The theoretical external quality factor is $Q_e = 67.4$. The coupling J_2 that controls the transmission zero is achieved using cross-coupling between resonators 1 and 3. The tuning of d_{13} will allow us to control the frequency of the transmission zero.

5.2.3 Defining the Dimensions of the Enclosure

The design technique relies on using simulations of the microwave structure to match the theoretical coupling coefficients of the prototype circuit. These simulations, however, require knowledge of the suspended substrate characteristics and the dimensions of the enclosure.

The enclosure acts as a cavity and TEM propagation modes can resonate. If the enclosure is not designed carefully, these resonances could seriously degrade the transmission and reflection characteristics of the filter. It is therefore important to choose the dimensions of this cavity carefully.

The resonance frequency of a TE_{mn} mode in a cavity of length L and width l is given by

$$f_{m,n} = \frac{c}{2\pi\sqrt{\varepsilon_{eq}}}\sqrt{\left(\frac{m\pi}{l}\right)^2 + \left(\frac{n\pi}{L}\right)^2}$$

where the various parameters used in this formula are as described in Chapter 3. Once the length L is known, it is possible to define the width l so that the enclosure-mode resonances do not occur in the operational frequency band of the filter. In addition, by using the length/width ratio $R = L/l$, the resonance frequency of a TE_{mn0} mode can be expressed as

$$f_{m,n} = \frac{c}{2L\sqrt{\varepsilon_{eq}}}\sqrt{n^2 + m^2R^2}$$

The cavity length L depends primarily on the value of the gap distance d_{13} that is used to obtain the theoretical coupling k_{13}. A first estimation of this coupling versus gap distance d_{13} without considering the enclosure provides a length $L = 16.68$ mm.

Using the estimation of the cavity length, the resonances of the enclosure versus the ratio R can be computed. Figure 5.4 shows the resonant frequencies of the modes in terms of R. When R is taken as 1.02, the first resonance occurs at 12.2 GHz, far beyond the filter passband centered at 7.28 GHz. Knowing L and R, the enclosure width can be calculated. In this case the cavity width should

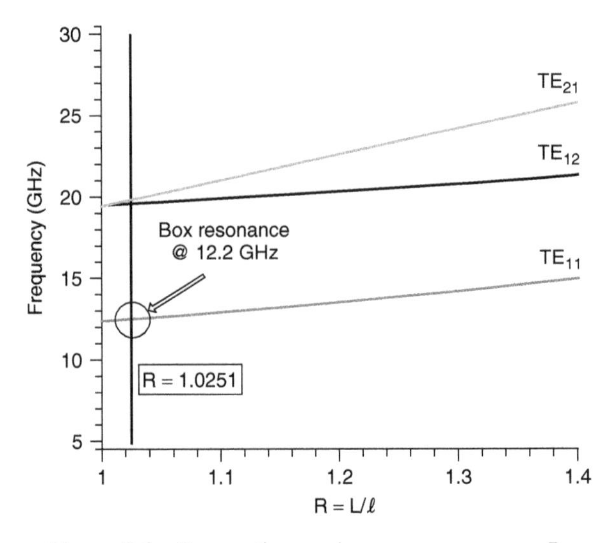

Figure 5.4 Propagation mode resonances versus R.

be $l = 16.27$ mm. When the cavity dimensions are known, simulations of the various elements of the filter can be used to match the theoretical couplings and external quality factor.

5.2.4 Defining the Gap Distances

The design of the filter starts with the coupling coefficient k_{13}, which controls the position of the transmission zero at 7.4 GHz. This coupling depends on d_{13}. Figure 5.5 shows the variations of the coupling k_{13} when d_{13} is changed. Using the figure we find that a gap distance $d_{13} = 2.92$ mm is needed to get the theoretical coupling $k_{13} = 0.0025$. The next step is to define the gap distance d_{12} to achieve the theoretical coupling k_{12}. Simulations showing the variations of the coupling coefficient k_{12} versus d_{12} are shown in Figure 5.6. A gap distance $d_{12} = 1.32$ mm is needed to get the theoretical coupling $k_{12} = 0.011$.

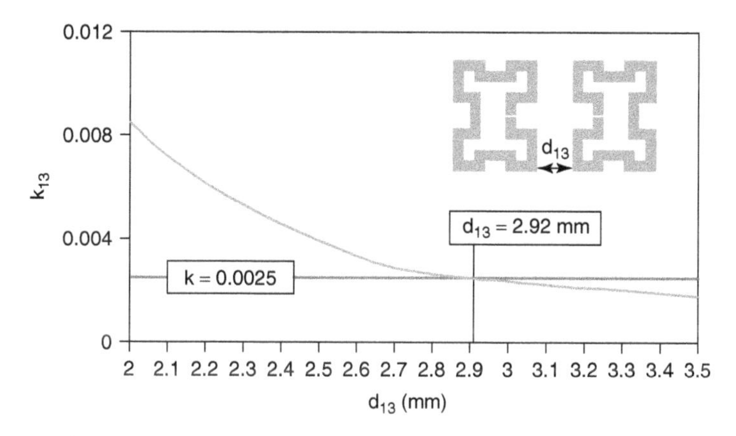

Figure 5.5 Variations of coupling k_{13} versus gap distance d_{13}.

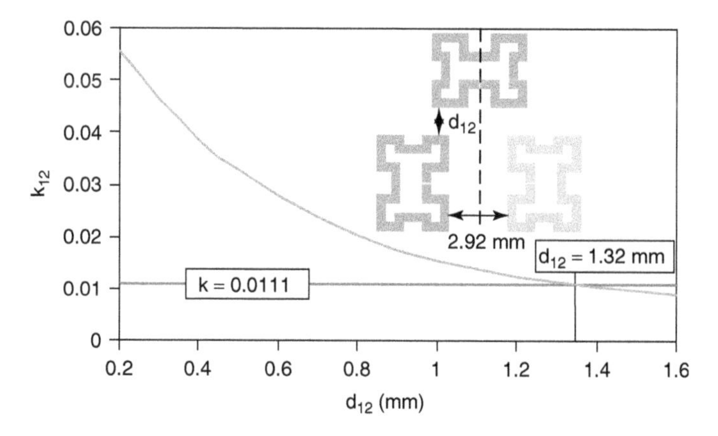

Figure 5.6 Variations of coupling k_{12} versus gap distance d_{12}.

The final step is to match the theoretical external quality factor of the filter Q_e. The external quality factor depends on the gap distance between the input line and resonator 1 (also the gap distance between resonator 3 and the output line). Figure 5.7 shows variations in Q_e when the distance between the input line and the first resonator changes. For a theoretical external quality factor of 67.4, a gap distance $d_{\text{in}} = 0.163$ mm is needed. The physical design parameters for the filter are summarized in Figure 5.8.

The propagation characteristics of the filter using the initial gap distances are shown in Figure 5.9. It can be seen that the actual couplings must be stronger than the theoretical couplings since the filter bandwidth is larger than expected, the return loss is less than 20 dB, and the transmission zero is shifted to higher frequencies. The gap distances must be adjusted using the complete structure of the filter instead of the isolated pieces used previously for defining suitable initial values. The final adjusted gap distances are given in Figure 5.8 and the

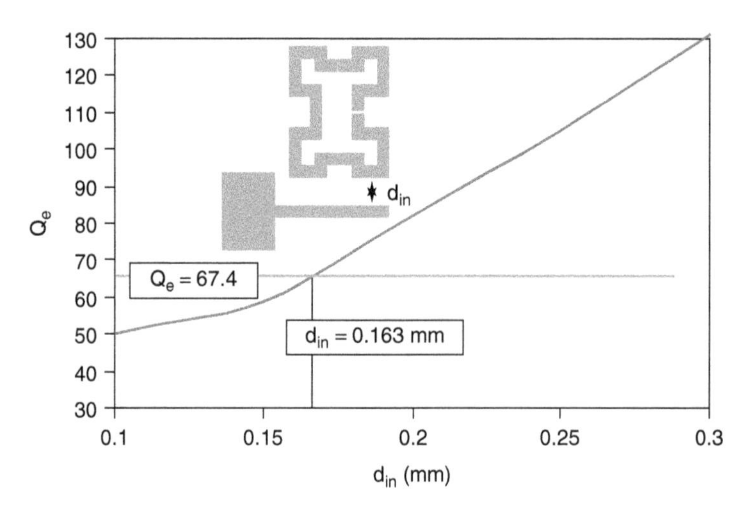

Figure 5.7 Variations of coupling Q_e versus gap distance d_{in}.

Initial dimensions	Dimensions after tuning
$d_{in} = d_{out} = 0.163$ mm	$d_{in} = d_{out} = 0.195$ mm
$d_{12} = d_{23} = 1.32$ mm	$d_{12} = d_{23} = 1.49$ mm
$d_{13} = 2.92$ mm	$d_{13} = 3.25$ mm

Figure 5.8 Initial and final gap distances.

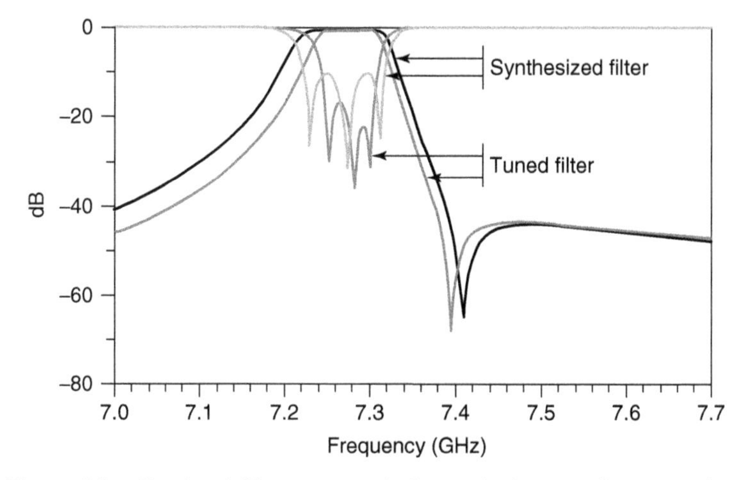

Figure 5.9 Simulated filter response before and after gap distance tuning.

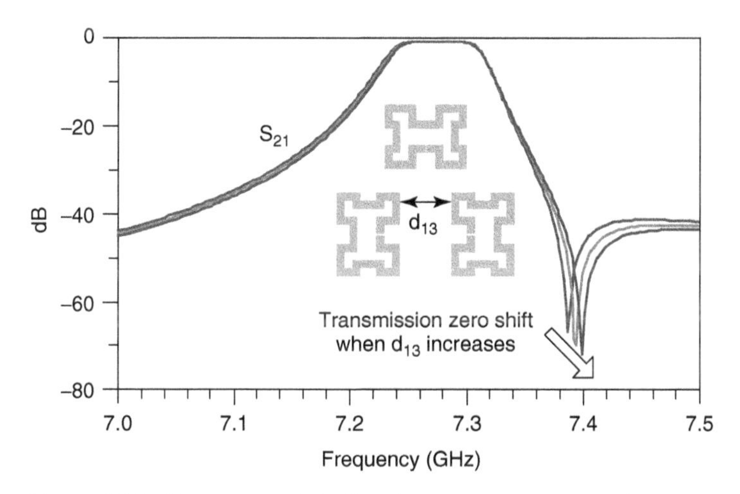

Figure 5.10 How d_{13} controls the placement of a transmission zero.

corresponding filter response in Figure 5.9. Figure 5.10 shows that the transmission zero can be placed accurately at 7.4 GHz by varying d_{13}. The coupling is magnetic (positive), as shown in the current distribution of Figure 5.11.

5.2.5 Further Study of the Filter Response

Figure 5.12 shows the propagation characteristics of the filter when metallic and dielectric losses are considered. The metallic losses are due to the circuit (copper) and shielding (brass), and the dielectric losses are due to the substrate. The in-band insertion loss reaches 6.3 dB, and the return loss is smaller than

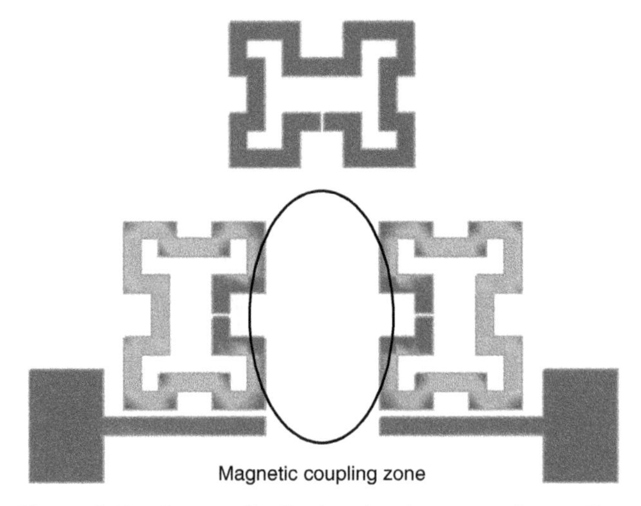

Figure 5.11 Current distribution showing magnetic coupling.

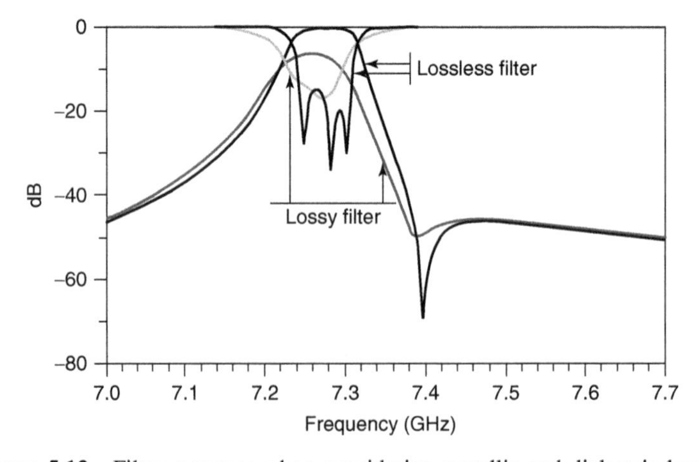

Figure 5.12 Filter response when considering metallic and dielectric losses.

17 dB. Figure 5.13 shows the wideband transmission characteristic of the filter. The resonant frequency of the enclosure is calculated by Agilent Momentum software at 12.5 GHz. The attenuation is greater than 50 dB between 7.5 and 12 GHz, and the attenuation at twice the center frequency is about 40 dB.

It is interesting to compare the filter responses of the meandered line filter with a traditional filter using straight $\lambda/2$ line resonators. The topology of the straight $\lambda/2$ line resonator filter is shown in Figure 5.14. The filter was designed using the same synthesis procedure as that presented in Chapters 2 and 3. Figure 5.15 shows the transmission characteristics of the two filters. It can be seen that the

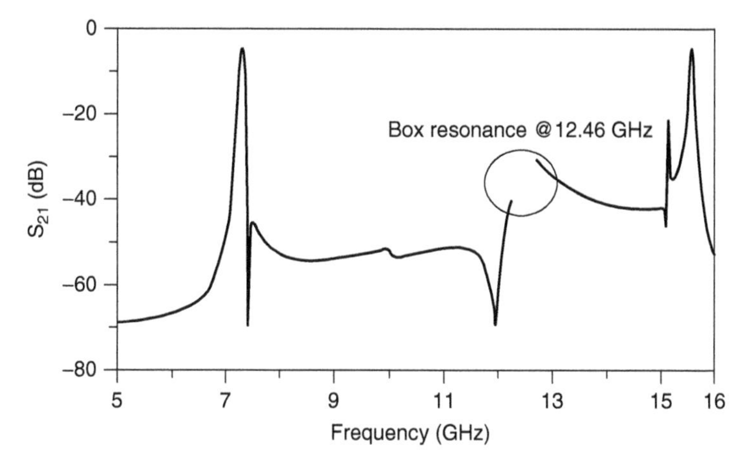

Figure 5.13 Simulated wideband transmission characteristic of the filter.

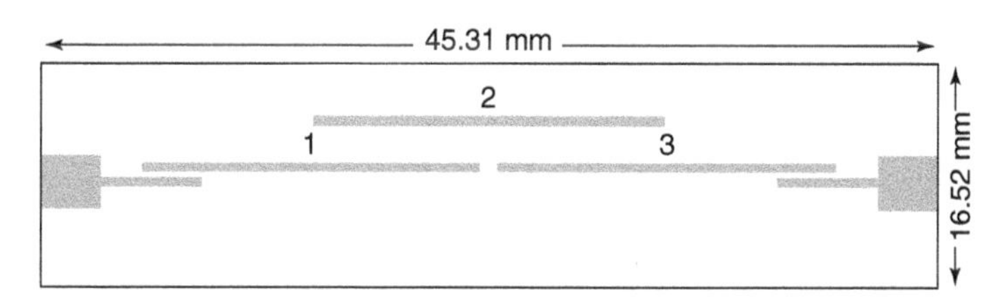

Figure 5.14 Topology of a third-order pseudo-elliptic $\lambda/2$ resonator filter.

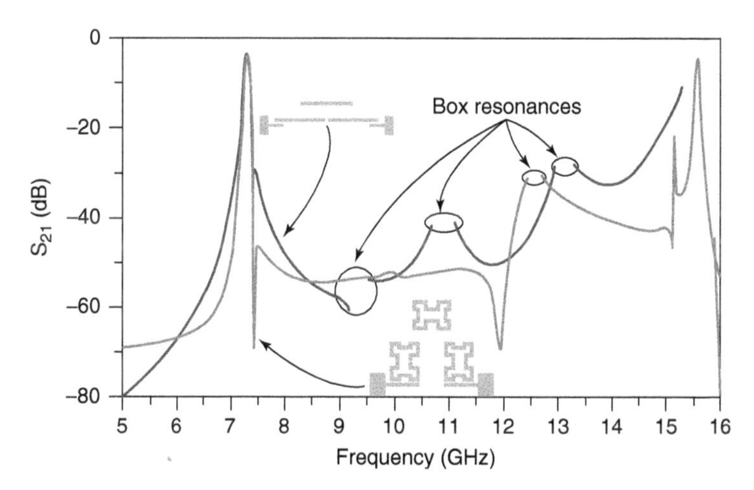

Figure 5.15 Meandered line and straight $\lambda/2$ line filter responses.

meandered line filter provides better out-of-band rejection, especially at the transmission zero frequency. The miniaturization process results in an enhancement of the rejection at twice the center frequency. The insertion loss is worse (i.e., 6.3 dB) for the meandered line filter than for the straight $\lambda/2$ line resonator's filter (i.e., 4.1 dB). However, the meandered line filter is significantly smaller (16 mm compared to 45 mm).

5.2.6 Realization and Measured Performance

The suspended substrate structure simulated by Momentum is shown in Figure 5.16(a). In practice, the enclosure has sidewall grooves that hold the substrate as shown in Figure 5.16(b). The presence of these grooves can have unexpected effects on the propagation characteristics of the filter if they are not dimensioned correctly. Figure 5.17 shows the effect of the

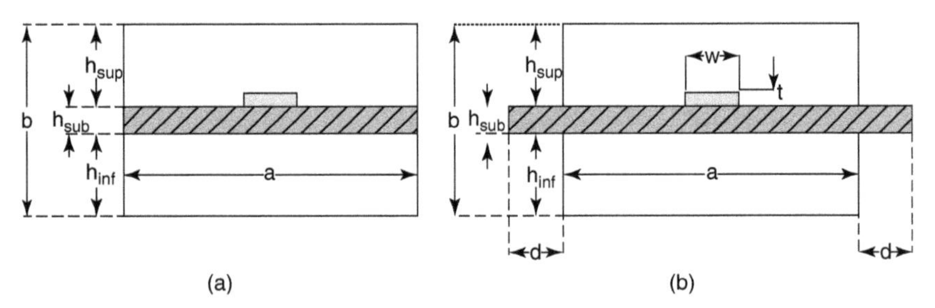

Figure 5.16 (a) Suspended substrate structure simulated by Momentum; (b) actual realization.

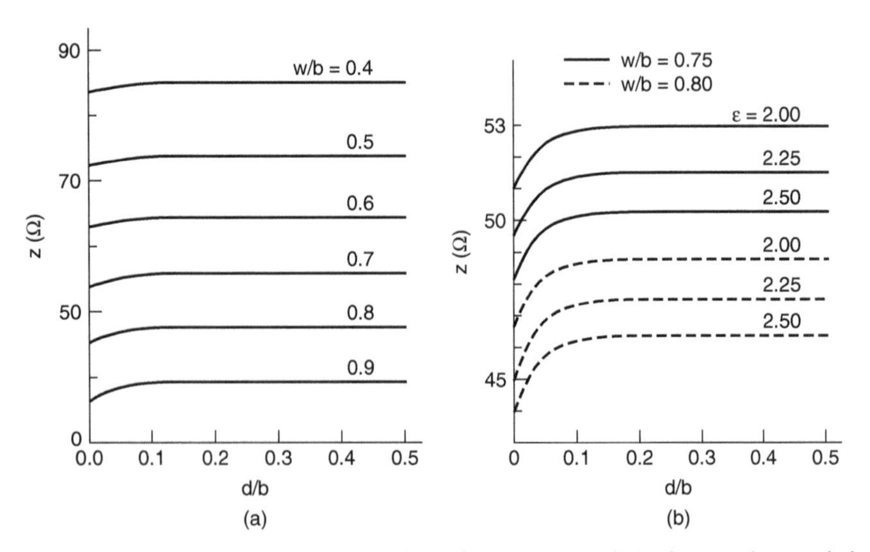

Figure 5.17 Effect of sidewall groove dimensions on suspended substrate characteristic impedance for (a) the Duroid substrate, and (b) other substrates.

grove dimensions on the characteristic impedance of the suspended substrate stripline.

The structure dimensions were chosen considering the most common characteristic impedance values, such as 50 Ω and 75 Ω. The substrate used is Duroid ($\varepsilon_r = 2.2$) and the dimensions are

$$\frac{a}{b} = 1 \qquad \frac{h_1}{b} = 0.4 \qquad \frac{h_2}{b} = 0.2 \qquad \frac{h_3}{b} = 0.4 \qquad t = 0.0$$

$$\frac{w}{b} = 0.2 - 0.9 \qquad \frac{d}{b} = 0.0 - 0.5$$

We notice that when the sidewall groove depth d increases, the characteristic impedance also increases, but this effect disappears for high values of d. To avoid undesirable effects, we must have $a \gg b$. One should also remember that a is important for calculation of the channel cutoff frequency. Moreover, we notice that the sidewall grooves don't affect the characteristic impedance when $d/b > 0.15$. In our case, $a = 16.27$ mm $\gg b = 2.254$ mm and the sidewall groove depth is $d = 1$ mm, giving $d/b = 0.44 > 0.15$. Hence, the conditions for a correct filter behavior are satisfied.

Major concerns when constructing a microwave filter are cost and complexity. The miniaturized filters will require small fabrication errors. When constructing rectangular cavities, exact square corners are difficult to obtain. For example, Figure 5.18 shows that the corners are more likely to be rounded, depending on the equipment available. The cavities were fabricated in a laboratory environment from blocks made of brass. Brass was selected since it is easier than copper to machine. On the other hand, it has greater losses. For cost reasons, copper was used for the suspended stripline [1]. The filters were assembled in the laboratory environment [2–10] and great caution was needed not to twist the thin flexible substrate.

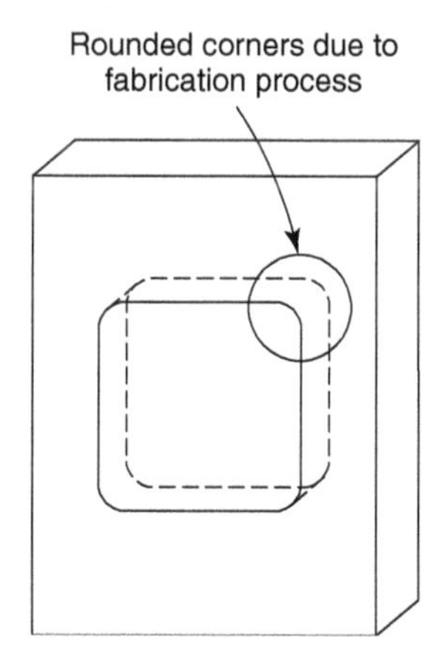

Figure 5.18 Fabricating a rectangular cavity.

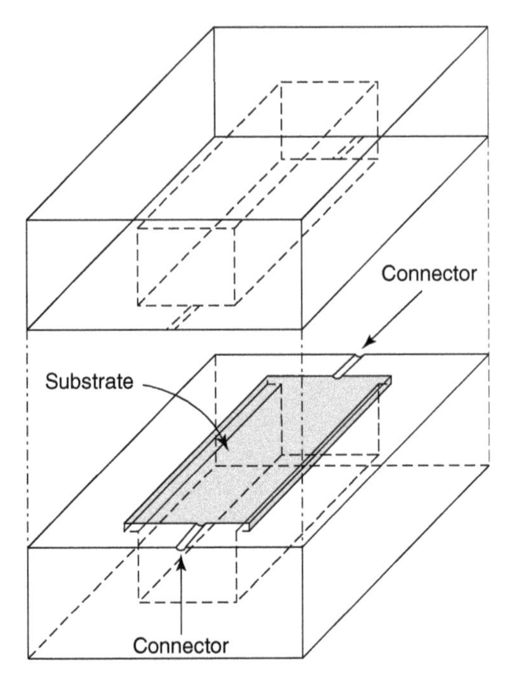

Figure 5.19 Overall view of the suspended substrate structure and enclosure.

Figure 5.19 shows a block of brass with the rectangular cavity cut in half to allow placing the suspended substrate. This introduces other imperfections, since the upper and lower cavities cannot be assembled perfectly, which results in perturbation of the electromagnetic propagation in the structure. To improve the insertion losses, the interior of the cavities could be covered with a thin layer of material with high electric conductivity, such as silver or gold.

The substrate and planar resonators (the filter) were fabricated by Atlantec, a general printed circuit board company which does not specialize particularly in microwave applications. When looking at the final product shown in Figure 5.20, it is clear that meandered line resonators do not appear to be particularly challenging to produce, and Atlantec did not report any major problems in fabricating them.

The performance of the filter was measured using network analyzer HP8720. Figure 5.21 shows the measured and simulated responses of the filter. The simulations include metallic and dielectric losses to facilitate comparison. Table 5.1 provides a set of key filter characteristics of the measured and simulated filter responses.

Figure 5.22 shows the simulated and measured wideband transmission characteristics of a meandered line filter. The measurements show a rejection superior to 48 dB between 5 and 7 GHz and a rejection superior to 40 dB between 7.5 and 12 GHz. At twice the center frequency the rejection is around 38 dB. The first spurious response appears at 15.8 GHz. The enclosure resonance occurs at 12.2 GHz instead of the expected 12.46 GHz.

Figure 5.20 Third-order meandered line pseudo-elliptic filter with a transmission zero on the right.

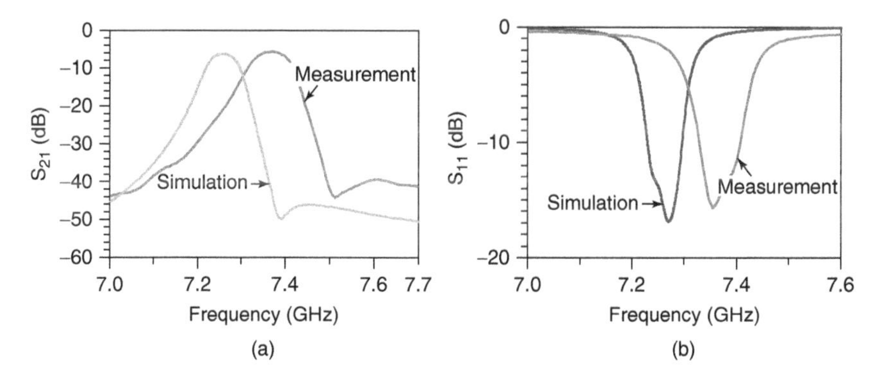

Figure 5.21 Measured and simulated filter responses.

TABLE 5.1 Key Measured and Simulated Meandered Line Filter Characteristics

Filter Characteristic	Simulation	Measurement
Central frequency	7.28 GHz	7.37 GHz
Bandwidth	100 MHz	112 MHz
Insertion loss	−6.3 dB	−6 dB
Return loss	≤ -17 dB	≤ -15 dB
Transmission zero	7.4 GHz	7.51 GHz

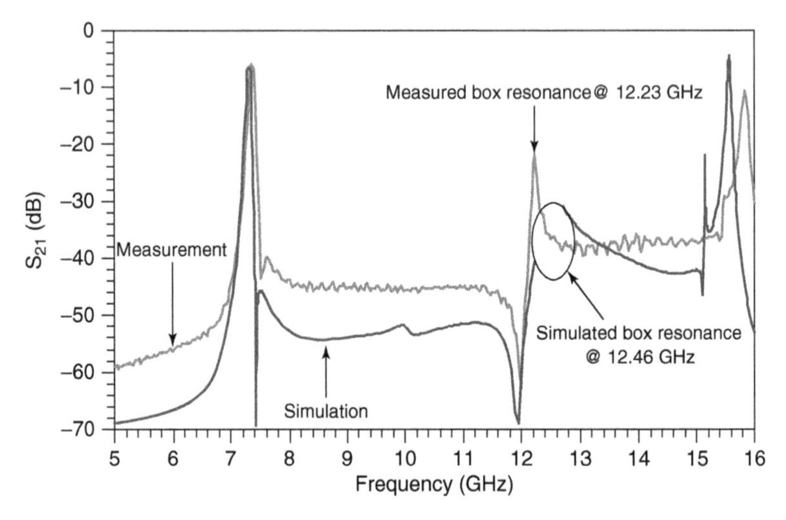

Figure 5.22 Measured and simulated transmission characteristics.

The measured central frequency is slightly higher than expected. This indicates that the couplings are stronger than expected. This can be due to cavity heights that are greater than specified (1 mm). It was seen in Chapter 4 that the coupling coefficient increases when the cavity height increases. The unloaded quality factor of the resonator also increases when the cavity height increases, reducing the insertion loss of the filter. This can explain why the measured insertion loss is smaller than expected. The measured enclosure resonance is lower than expected. This also indicates that the cavity heights might be greater than expected.

5.2.7 Sensitivity Analysis

We simulated the filter when the upper and lower cavity heights were changed by 0.1 mm. The results are reported with the measured response of the filter in Figure 5.23. When the cavity heights increase, the filter characteristics are shifted toward higher frequencies and the filter bandwidth increases due to an increase in coupling coefficients. The insertion loss decreases when the cavity heights increase, due to an increase in the unloaded quality factor of the resonators.

We also simulated the filter when the substrate thickness was changed by 0.04 mm. The cavity heights were kept at 1 mm. Figure 5.24 shows the results with the measured response of the filter. When the substrate thickness decreases, the filter propagation characteristics are shifted toward higher frequencies. For an electromagnetic simulation, the Momentum software decomposes the structure into elementary triangles or rectangles. It then calculates the propagation characteristics in each of these cells before generating the propagation characteristics of the entire structure. The mesh density should therefore be increased to achieve more accurate simulation results. On the other hand, the simulation time is increased strongly when the mesh density is increased. The mesh density

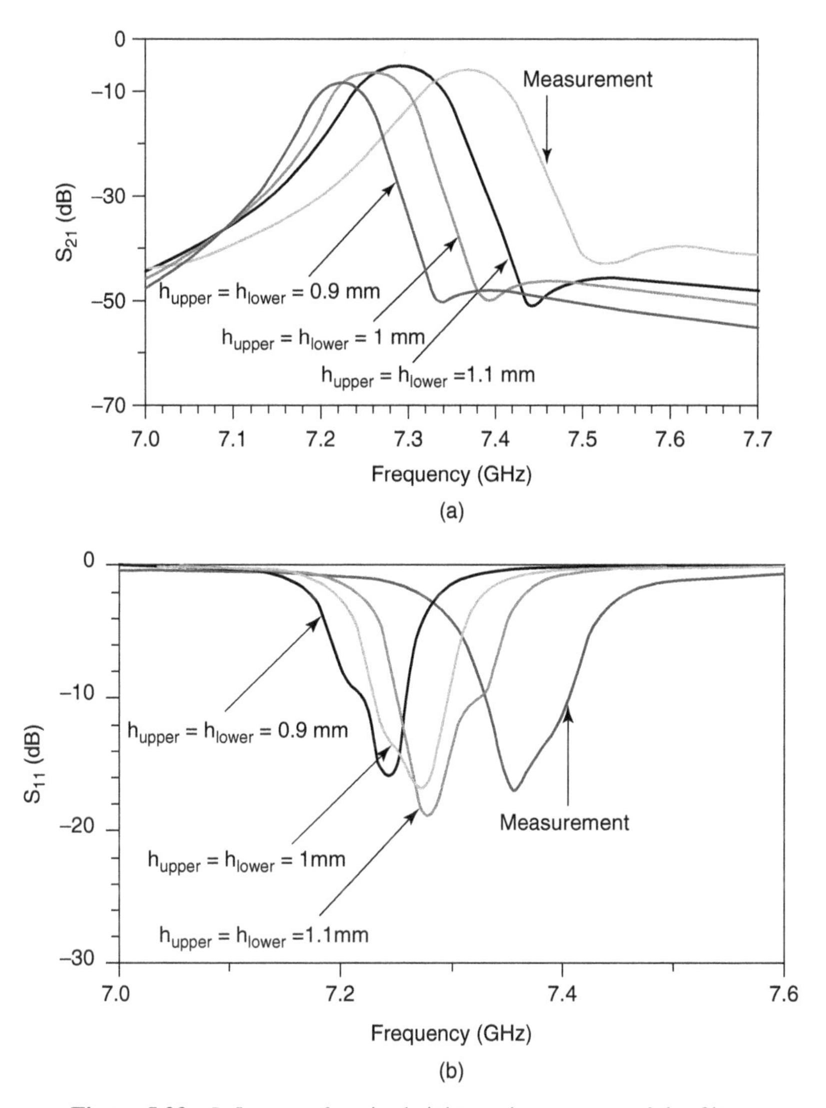

Figure 5.23 Influence of cavity height on the response of the filter.

is expressed in the number of cells by wavelength. In general, a mesh density of 30 cells/λ is considered as an acceptable compromise between precision and simulation time. We simulated the filter using mesh densities of 40 and 50 cells/λ. These simulations were carried out using a computer with a Pentium IV processor with 1 GB of RAM capacity. Figure 5.25 shows the simulation results and corresponding simulation times. These results show that when the mesh density increases, the propagation characteristics are shifted toward higher frequencies. Thus, it is not surprising that compared to simulations, the measured results are shifted to higher frequencies.

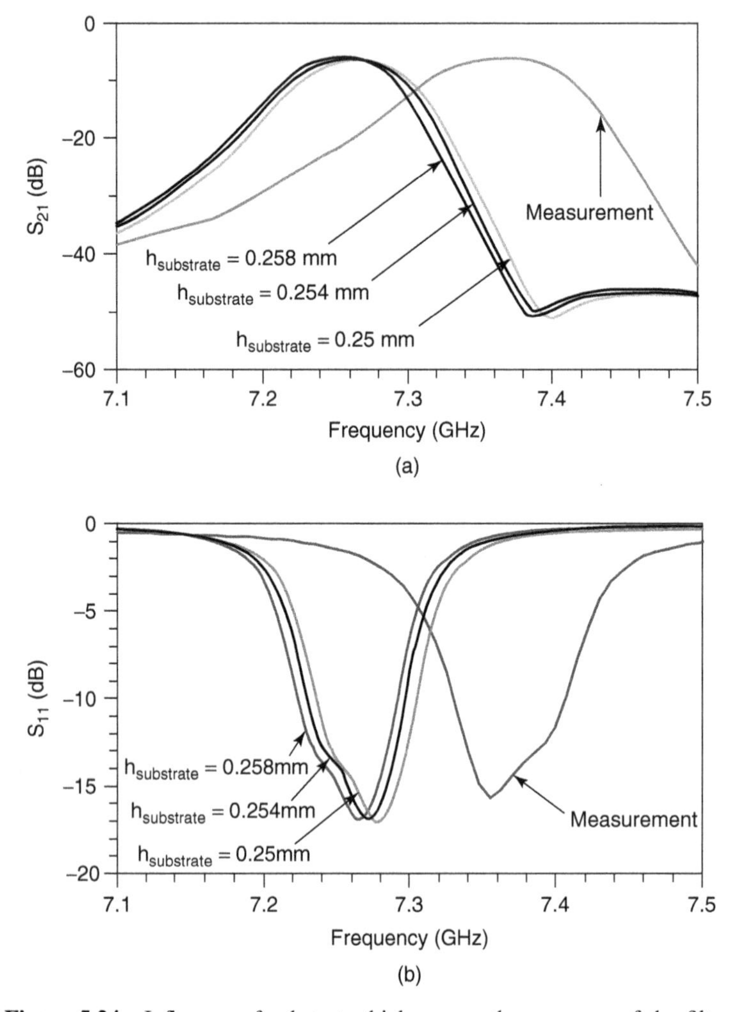

Figure 5.24 Influence of substrate thickness on the response of the filter.

In conclusion, the measurement results show good agreement with the simulations even though they are generally shifted toward higher frequencies. Using additional simulation results we can suppose that the substrate was thinner than specified, and the cavities were higher than expected. For precise simulation results, a high mesh density is required, even if the simulation time is increased. Moreover, the SMA connectors introduce supplementary losses. However, the out-of-band rejection and the selectivity obtained are very good for a third-order filter. The first spurious response occurs beyond twice the center frequency of the filter. The insertion losses could be reduced by using a better-conducting material than copper or by using a different technology that provides higher quality factors, such as superconductivity or micromachining.

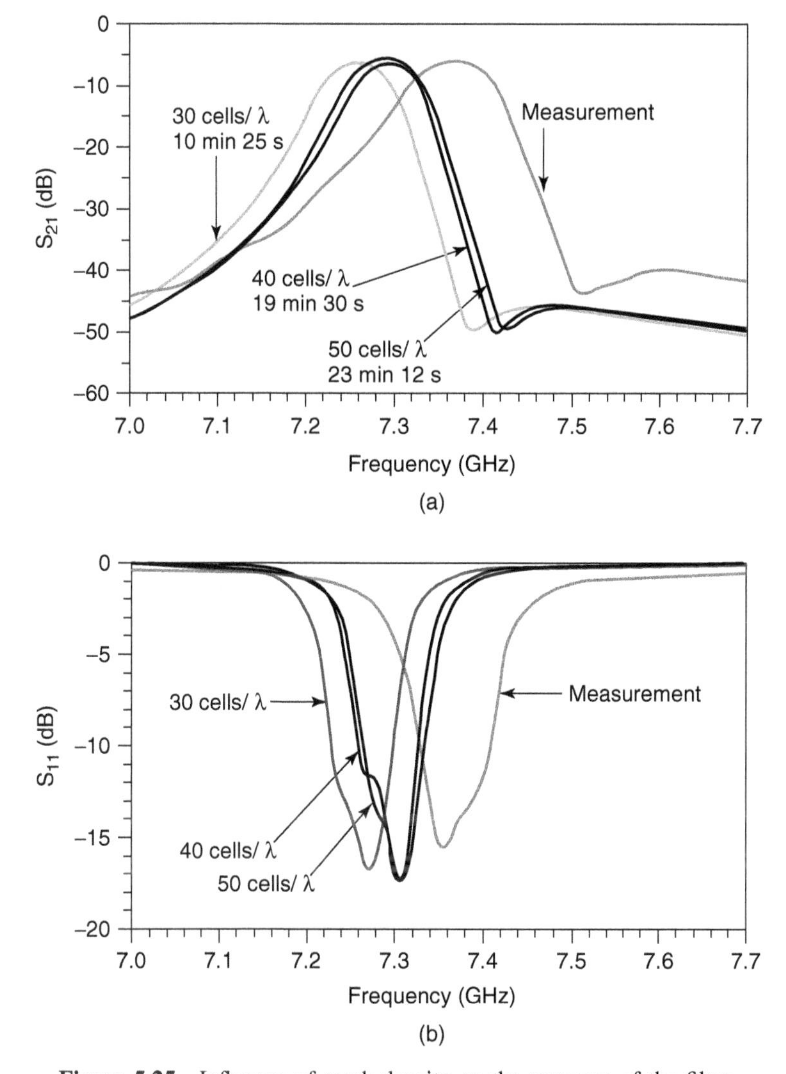

Figure 5.25 Influence of mesh density on the response of the filter.

5.3 THIRD-ORDER PSEUDO-ELLIPTIC FILTERS WITH TRANSMISSION ZERO ON THE LEFT

5.3.1 Topology of the Filter

The topology of a third-order pseudo-elliptic filter with a transmission zero on the left side of the passband is presented in Figure 5.26. The filter uses the same topology as before except that the resonators have been flipped to achieve an electrical coupling between resonators 1 and 3 to create a transmission zero on the left side of the passband.

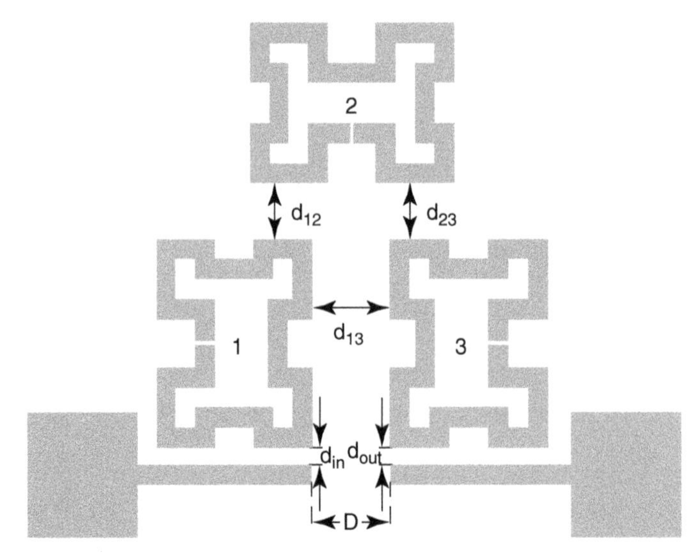

Figure 5.26 Topology of the third-order pseudo-elliptic filter with a transmission zero on the left.

5.3.2 Synthesis of the Bandpass Prototype

The design of the filter starts with the synthesis of the bandpass pseudo-elliptic prototype circuit capable of producing the desired filtering characteristics. For example, we are interested in designing a third-order pseudo-elliptic filter centered at 7.28 GHz with a bandwidth of 100 MHz (1.37%) and having a transmission zero on the left side of the passband at 7.2 GHz. Figure 5.27 shows the lowpass and bandpass third-order pseudo-elliptic prototypes and the bandpass transmission and reflection responses. The theoretical couplings needed for the design of the filter are $K_{12} = 0.0134$, $K_{13} = -0.0066$, and $K_{23} = 0.0134$. The theoretical external quality factor is $Q_e = 67.4$. The negative coupling J_2 that controls the location of the transmission zero is achieved using cross-coupling between resonators 1 and 3. Tuning d_{13} makes it possible to control the transmission zero frequency.

5.3.3 Defining the Dimensions of the Enclosure

Due to the similarities of the filter characteristics, the same suspended substrate enclosure is used. The enclosure has a length of 16.68 mm and a width of 16.27 mm. The first enclosure resonance occurs at 12.2 GHz.

5.3.4 Defining the Gap Distances

We start with the coupling coefficient k_{13} for placement of the transmission zero on the left of the passband at 7.2 GHz. This coupling depends on d_{13}. According

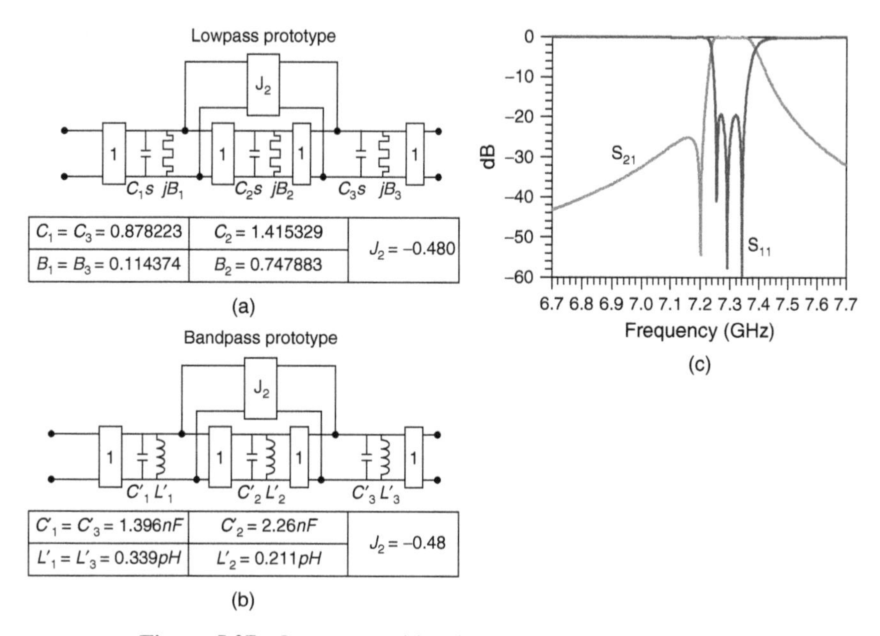

Figure 5.27 Lowpass and bandpass prototypes for the filter.

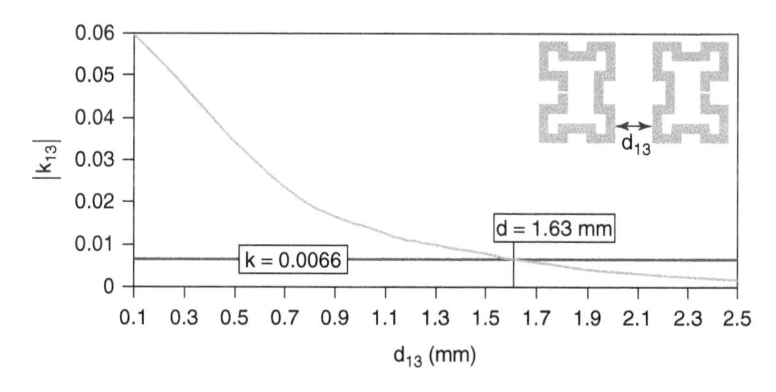

Figure 5.28 Variations of coupling coefficient k_{13} versus gap distance d_{13}.

to the coupling coefficient matrix, we must have $k_{13} = -0.0066$. A negative value for the coupling is assured by the electric coupling between resonators 1 and 3. Figure 5.28 shows the variations of k_{13} with d_{13}. According to the figure, $|k_{13}| = 0.0066$ when $d_{13} = 1.63$ mm.

When the gap spacing between resonators 1 and 3 is defined, the variation of k_{12} versus d_{12} is used to match the next theoretical coupling. Figure 5.29 shows variations in the coupling coefficient k_{12} as a function of the gap between resonators 1 and 2. According to the figure, for $k_{12} = 0.0134$ we find that $d_{12} = 1.29$ mm. From symmetry we will also have $k_{13} = 0.0134$ and $d_{13} = 1.29$ mm.

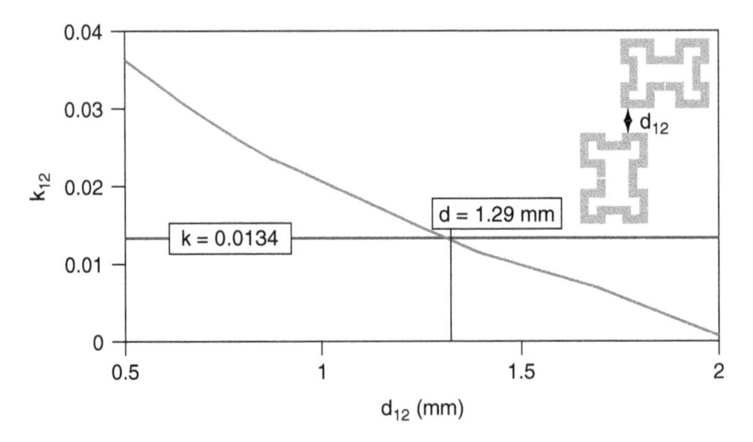

Figure 5.29 Variations of coupling coefficient k_{12} versus gap distance d_{12}.

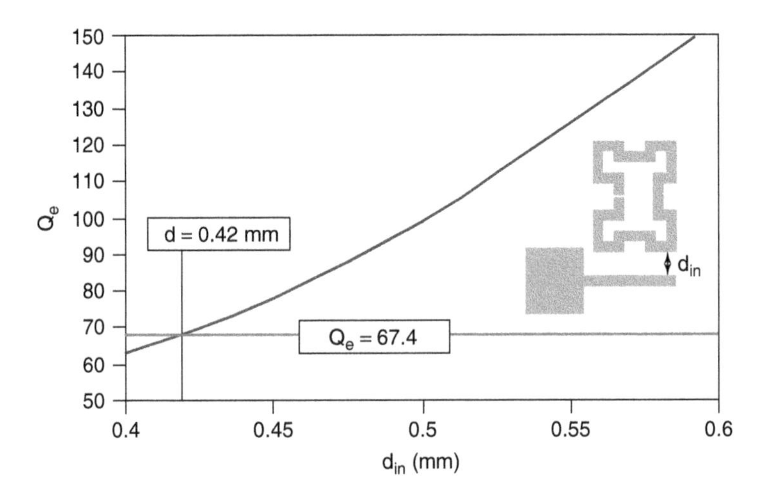

Figure 5.30 Variations of the external quality factor versus gap distance d_{in}.

Finally, the external quality factor is defined by the gap spacing between the input line and resonator 1. Figure 5.30 shows variations in the external quality factor Q_e versus gap distance d_{in}. For the theoretical value $Q_e = 67.4$, we should use $d_{\text{in}} = 0.42$ mm. The filter with the initial gap distances is shown in Figure 5.31, and the response of the filter using the initial gap distances is shown in Figure 5.32. The simulated transmission characteristic shows the presence of two supplementary transmission zeros, one at 6.75 GHz and the other at 7.87 GHz.

Figure 5.32 shows that the bandwidth is larger than expected, indicating that the effective couplings must be stronger than expected. By using simulations of the entire filter rather than the isolated simulations used for defining the initial gap distances, it is possible to readjust the coupling coefficients. The gap values after

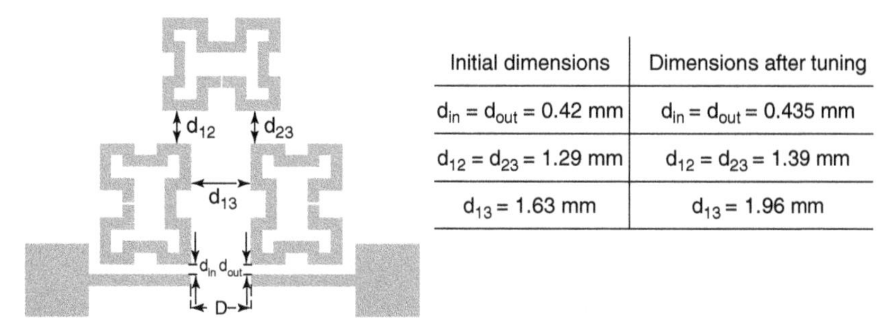

Initial dimensions	Dimensions after tuning
$d_{in} = d_{out} = 0.42$ mm	$d_{in} = d_{out} = 0.435$ mm
$d_{12} = d_{23} = 1.29$ mm	$d_{12} = d_{23} = 1.39$ mm
$d_{13} = 1.63$ mm	$d_{13} = 1.96$ mm

Figure 5.31 Initial and final gap distances.

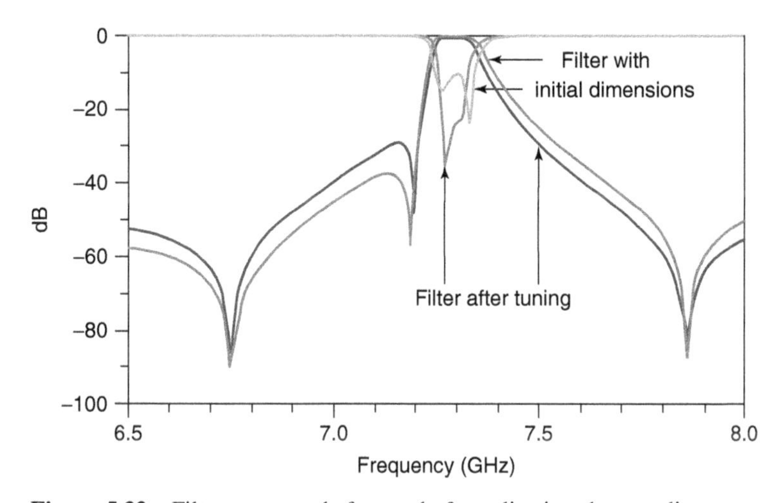

Figure 5.32 Filter response before and after adjusting the gap distances.

final adjustments are reported in Figure 5.31, and the corresponding filter response is given in Figure 5.32. The coupling that required the greatest adjustment was k_{13}. It is influenced by the coupling between the input and output lines (distance D). This coupling increases the value k_{13}. Therefore, d_{13} must be increased to recover the desired value of k_{13}. Moreover, the coupling between the feed lines creates two additional transmission zeros, one at 6.75 GHz and the other at 7.87 GHz. Figure 5.33 shows the current distribution at these two frequencies and allows us to see the resonances of the feedlines. It also shows how D can move the two transmission zeros without changing the prescribed transmission zero at 7.2 GHz. The zero at 7.2 GHz is due to the electric coupling between resonators 1 and 3, as shown in Figure 5.34(a). Adjusting the gap distance d_{13} makes it possible to change the coupling coefficient k_{13} and to position the prescribed transmission zero correctly, as shown in Figure 5.34(b).

Figure 5.35 shows the degradation in filter response when dielectric and metallic losses are considered. The in-band insertion loss reaches 4.3 dB, and the return loss drops below 20 dB.

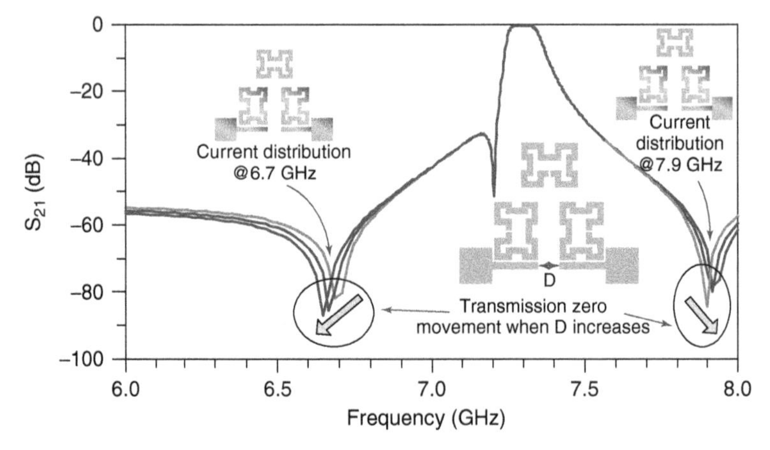

Figure 5.33 Transmission zeros due to the coupling between feed lines.

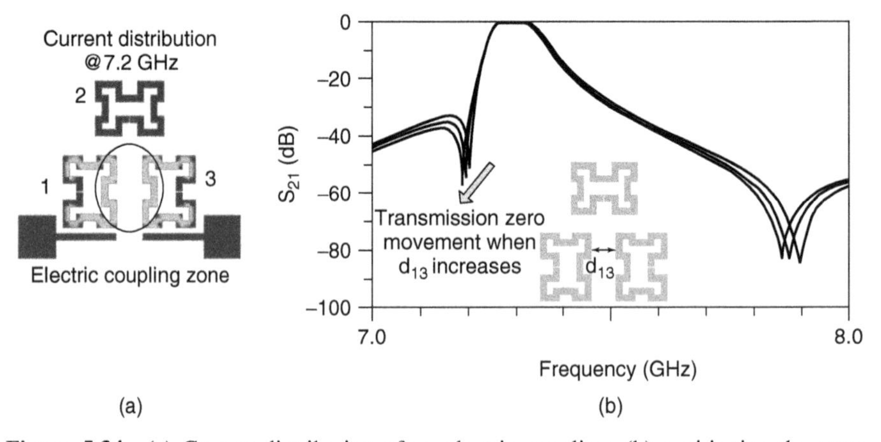

Figure 5.34 (a) Current distribution of an electric coupling; (b) positioning the transmission zero using gap distance d_{13}.

5.3.5 Realization and Measured Performance

The fabrication followed the same procedures as in Section 5.2.6. The enclosure was built in-house and the substrate and planar resonators (the filter) were fabricated by Atlantec. The constructed third-order pseudo-elliptic filter with a transmission zero on the left of the passband is shown in Figure 5.36. The simulated and measured responses are given in Figure 5.37. Table 5.2 provides some of the key measured and simulated characteristics of the filter.

Figure 5.38 provides the simulated and measured wideband transmission characteristic of the filter. The measured rejection is greater than 48 dB between 5 and 7 GHz, and the measured rejection is greater than 40 dB between 7.5 and 12 GHz. The measured rejection is around 38 dB at twice the center frequency of

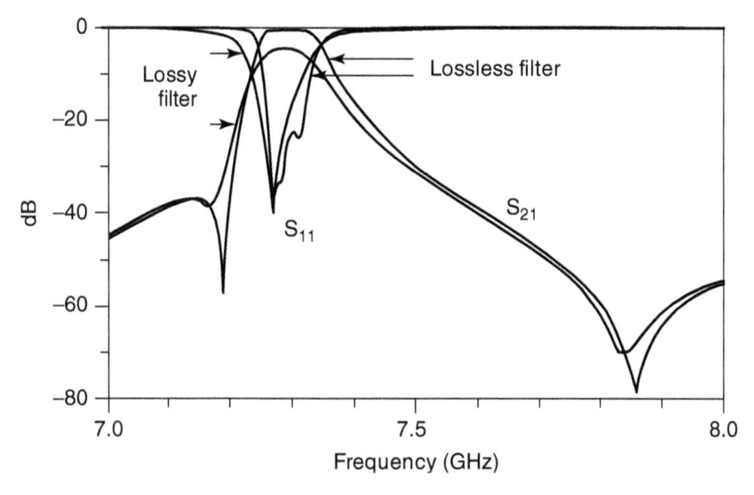

Figure 5.35 Effects of metallic and dielectric losses on the response of the filter.

Figure 5.36 Third-order meandered line pseudo-elliptic filter with a transmission zero on the left.

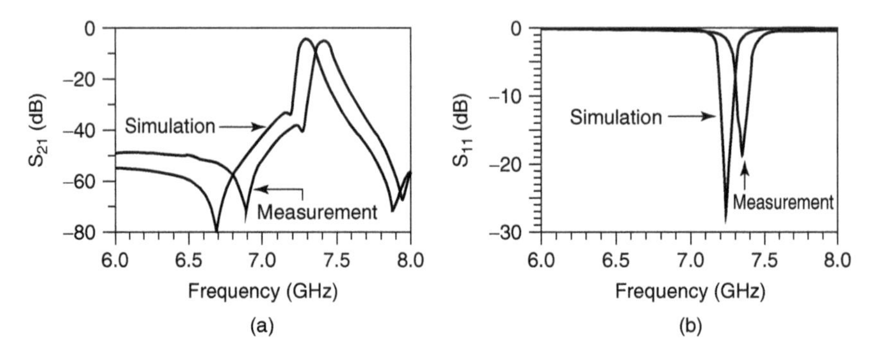

Figure 5.37 Simulated and measured responses of the filter.

TABLE 5.2 Key Measured and Simulated Filter Characteristics

Filter Characteristic	Simulation	Measurement
Central frequency	7.28 GHz	7.37 GHz
Bandwidth	98 MHz	105 MHz
Insertion loss	−4.3 dB	−5 dB
Return loss	≤ −20 dB	≤ −13 dB
Transmission zero	7.18 GHz	7.26 GHz

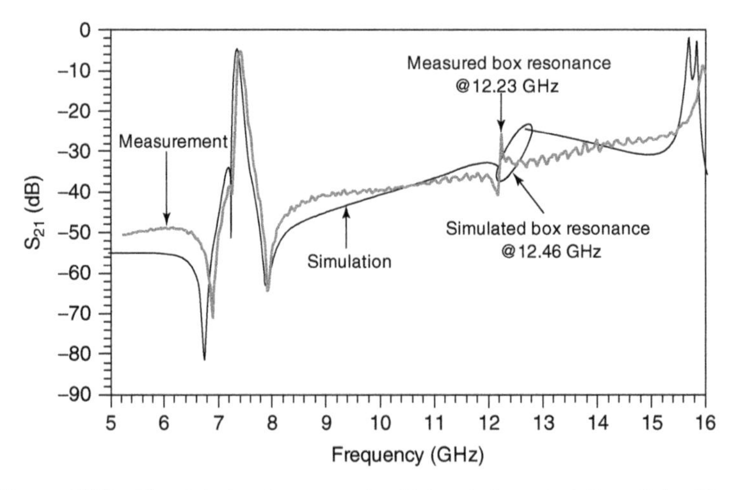

Figure 5.38 Simulated and measured wideband characteristics of the filter.

the filter. The first spurious response measured occurs at 15.8 GHz. The enclosure resonance is measured at 12.23 GHz compared to the 12.46 GHz expected. The measured center frequency is higher than expected. This indicates that the couplings are stronger than specified. As for the case of the previous filter, additional simulations are conducted to check the effects of fabrication tolerances.

5.3.6 Sensitivity Analysis

First, the upper and lower cavity heights are changed by 0.1 mm. The simulation results and the measured response are reported in Figure 5.39. As for the previous filter, when the cavity heights increase, the filter characteristics are shifted to higher frequencies. The filter bandwidth increases due to the increase in coupling coefficients. The insertion loss decreases when the cavity heights increase, due to an improvement in the unloaded quality factor of the resonators.

The substrate thickness is then changed by 0.04 mm. The cavity heights are fixed to 1 mm. Figure 5.40 shows the effects of substrate thickness on the propagation characteristics of the filter. When the substrate thickness decreases, the

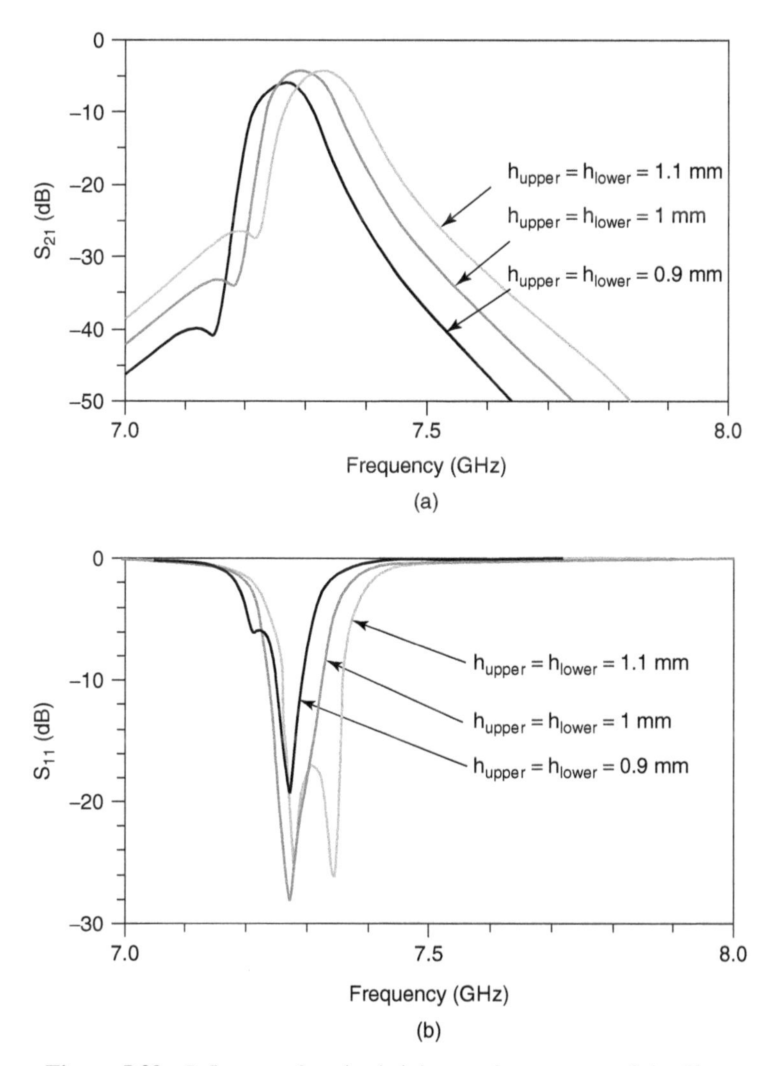

Figure 5.39 Influence of cavity heights on the response of the filter.

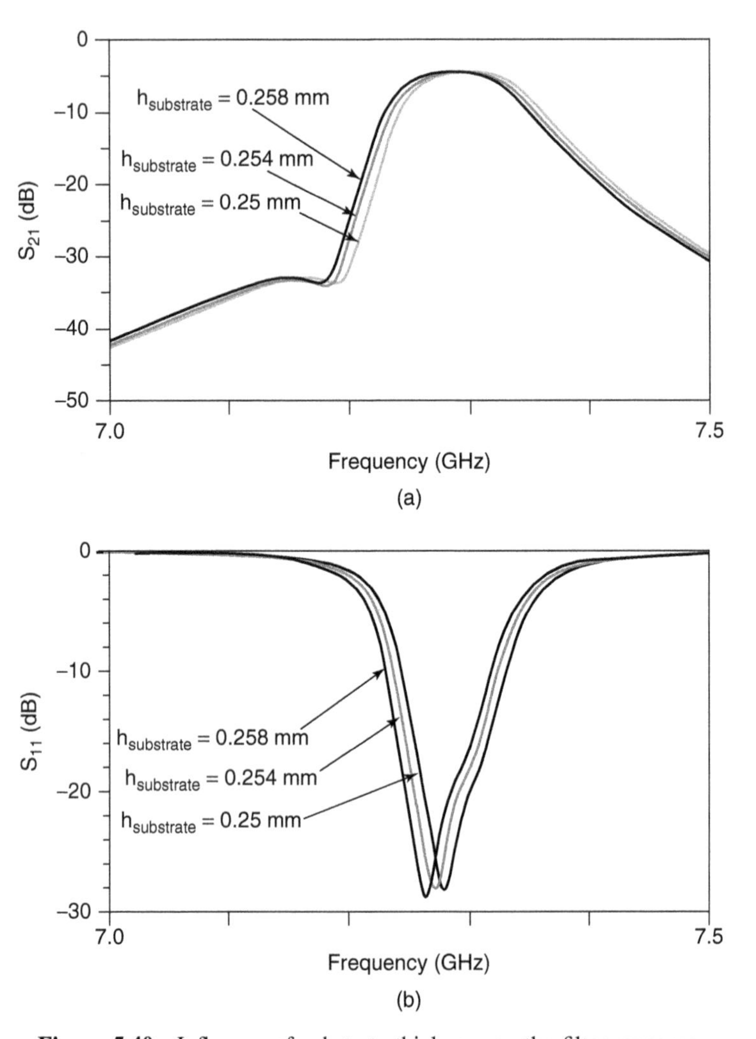

Figure 5.40 Influence of substrate thickness on the filter response.

filter propagation characteristics are shifted toward higher frequencies. Finally, Figure 5.41 shows the effects of mesh density on the propagation characteristics of the filter for mesh densities of 30 cells/λ, 40 cells/λ, and 50 cells/λ. When the mesh density increases, the filter characteristics are shifted toward higher frequencies.

In conclusion, as for the previous filter, measurement results show good agreement with the simulations even if they are shifted toward higher frequencies. From the additional simulations we can think that the actual substrate is thinner than expected and the actual cavities are higher than expected. For precise simulation results, a high mesh density is required, even if the simulation time is increased. Moreover, the SMA connectors introduce supplementary losses. The

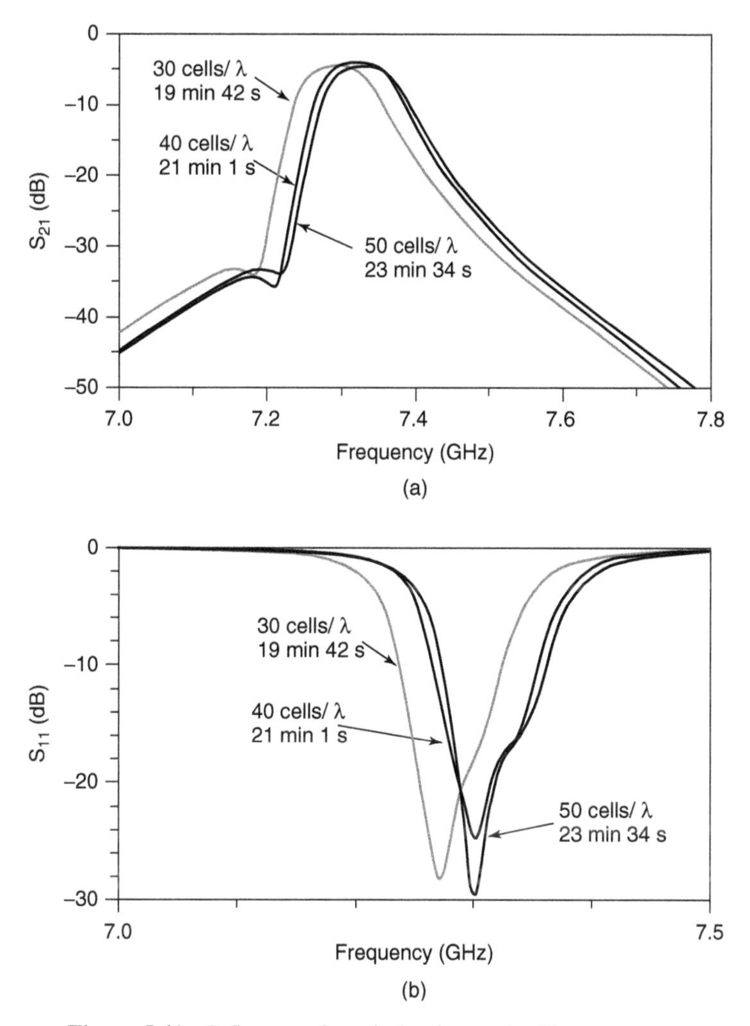

Figure 5.41 Influence of mesh density on the filter response.

out-of-band rejection and selectivity obtained are very good for a third-order filter. The first spurious response is beyond twice the center frequency of the filter, and the insertion and return losses could be reduced by using better conducting materials than copper or by using technologies with higher quality factors, such as superconductivity and micromachining.

REFERENCES

[1] I. J. Bahl and P. Bhartia, *Microwave Solid State Circuit Design*, Sec. 6.4., Wiley Hoboken, NJ, 2004.

[2] E. Hanna, P. Jarry, E. Kerhervé, J. M. Pham, and D. Lo Hine Tong, Synthesis and design of a suspended substrate capacitive-gap parallel coupled-line (CGCL) bandpass filter with one transmission zero, *10th IEEE International Conference on Electronics Circuits and Systems (ICECS 2003)*, University of Sharjah, United Arab Emirates, pp. 547–550, Dec. 14–17, 2003.

[3] E. Hanna, P. Jarry, E. Kerhervé, and J. M. Pham, General prototype network method for suspended substrate microwave filters with asymmetrical prescribed transmission zeros: synthesis and realization, *Mediterranean Microwave Symposium (MMS 2004)*, Marseille, France, pp. 40–43, June 1–3, 2004.

[4] E. Hanna, P. Jarry, E. Kerhervé, and J. M. Pham, General prototype network with cross coupling synthesis method for microwave filters: application to suspended substrate strip line filters, *ESA–CNES International Workshop on Microwave Filters*, Toulouse, France, Sept. 13–15, 2004.

[5] E. Hanna, P. Jarry, E. Kerhervé, and J. M. Pham, Etude de résonateurs planaires fractals et application à la synthèse de filtres microondes en substrat suspendu, *14th National Microwave Days (JNM 2005)*, Nantes, France, May 11–13, 2005, No. 7A4, Abstract, p. 134.

[6] E. Hanna, P. Jarry, E. Kerhervé, and J. M. Pham, Suspended substrate stripline cross-coupled trisection filters with asymmetrical frequency characteristics, *35th European Microwave Conference (EuMC)*, Paris, pp. 273–276, June 4–6, 2005.

[7] E. Hanna, P. Jarry, E. Kerhervé, and J. M. Pham, Cross-coupled suspended stripline trisection bandpass filters with open-loop resonators, *International Microwave and Optoelectronics Conference (IMOC2005)*, Brasília, Brazil, pp. 42–46, July 2005.

[8] E. Hanna, P. Jarry, E. Kerhervé, and J. M. Pham, Novel compact suspended substrate stripline trisection filters in the Ku band, *2005 Mediterranean Microwave Symposium (MMS2005)*, Athens, Greece, pp. 146–149, Sept. 6–8, 2005.

[9] E. Hanna, P. Jarry, E. Kerhervé, and J. M. Pham, A design approach for capacitive gap parallel-coupled line filters and fractal filters with the suspended substrate technology, in *Microwave Filters and Amplifiers* (scientific editor P. Jarry), Research Signpost, Tivandrum, India, pp. 49–71, 2005.

[10] J. M. Pham, P. Jarry, E. Kerhervé, and E. Hanna, A novel compact suspended stripline band pass filter using open-loop resonators technology with transmission zeroes, *International Microwave and Optoelectronics Conference (IMOC 2007)*, Salvador, Brazil, pp. 937–940, Oct. 2007.

CHAPTER 6

DESIGN AND REALIZATIONS OF HILBERT FILTERS

6.1 Introduction 129
6.2 Design of hilbert filters 130
6.2.1 Second-order Chebyshev responses 130
6.2.2 Third-order Chebyshev responses 134
6.2.3 Third-order pseudo-elliptic responses 136
6.2.4 Third-order pseudo-elliptic responses using second-iteration resonators 144
6.3 Realizations and measured performance 148
6.3.1 Third-order pseudo-elliptic filter with transmission zero on the right 149
6.3.2 Third-order pseudo-elliptic filter with transmission zero on the left 154
References 159

6.1 INTRODUCTION

The conception of filters using Hilbert resonators starts with the synthesis of the equivalent lowpass and bandpass prototype networks presented in Chapter 2. The synthesis ends by computing the theoretical elements of the coupling matrix and external quality factors. The theoretical coupling coefficients are used to define

Design and Realizations of Miniaturized Fractal RF and Microwave Filters,
By Pierre Jarry and Jacques Beneat

the gap distance between Hilbert resonators. The theoretical external quality factor is used to define the couplings between the input/output lines and the first/last Hilbert resonators.

6.2 DESIGN OF HILBERT FILTERS

6.2.1 Second-Order Chebyshev Responses

The Hilbert filter topology to provide a second-order Chebyshev response is shown in Figure 6.1. The filter is made of two first-iteration Hilbert resonators and the input and output lines. The design technique starts by computing the elements of the bandpass prototype to have the desired filtering characteristics using the synthesis method of Chapter 2. Figure 6.2 shows the bandpass prototype and the bandpass transmission and reflection responses of a second-order Chebyshev

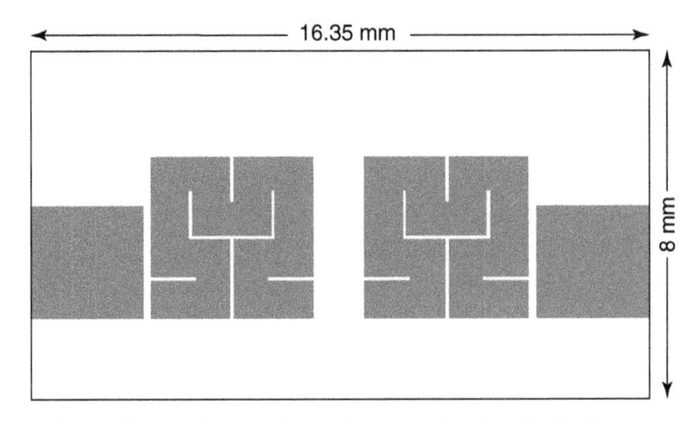

Figure 6.1 Hilbert filter topology for second-order Chebyshev responses.

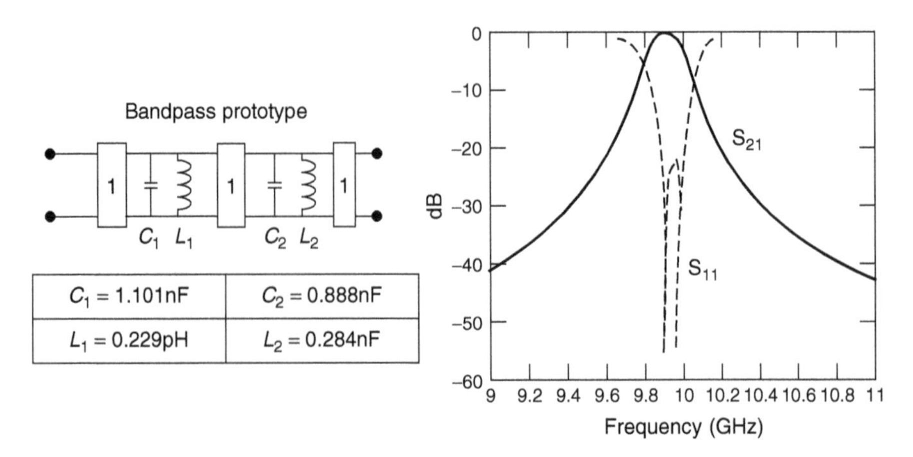

Figure 6.2 Second-order Chebyshev bandpass prototype.

filter centered at 9.9 GHz with a 80-MHz bandwidth. The theoretical couplings needed for the design of the filter are $K_{12} = 0.013$ and the theoretical external quality factor is $Q_e = 101.2$.

The design technique is based strongly on using computer simulations to define the various physical parameters of the filter, such as the gaps between resonators and between input/output lines and first/last Hilbert resonators. To use simulations, one must first define the suspended substrate and enclosure characteristics. For example, we will use a Duroid substrate with a relative dielectric permittivity of 2.2 with a thickness of 0.254 mm. The upper and lower cavity heights will be taken as 1 mm. The width of the input/output lines is computed using Linecalc from Agilent to provide 50 Ω of characteristic impedance. In this case the width of the input/output lines is found to be 3.11 mm. The resonators for 9.9 GHz have a side of 4.3 mm. The width of the enclosure is taken as 8 mm, and a first estimation of the gaps suggests that an enclosure length of 16.35 mm should be adequate. Using these parameters, the enclosure resonance occurs at 20.2 GHz, about twice the center frequency of the desired filter.

First, the distance between the input line and the first resonator is defined. Figure 6.3 shows the simulation resulting from computing coupling between the input line and the first resonator when the distance d_{in} is varied. The coupling between the input line and first resonator is matched to the theoretical external quality factor $Q_e = 101.02$, leading to $d_{\text{in}} = 0.182$ mm. Then the coupling between resonators 1 and 2 is computed for differing values of the gap distance d_{12}, as shown in Figure 6.4. The simulated coupling is then matched to the theoretical coupling coefficient $k_{12} = 0.013$. One finds that the distance between resonators should be $d_{12} = 1.26$ mm.

Simulations of the complete filter show that when everything is placed together, the couplings are stronger than expected. The gaps are then adjusted manually to recover the desired filtering characteristics. The final filter is shown in Figure 6.5. Figure 6.6 shows the transmission and reflection characteristics of the filter with and without taking dielectric and metallic losses into account.

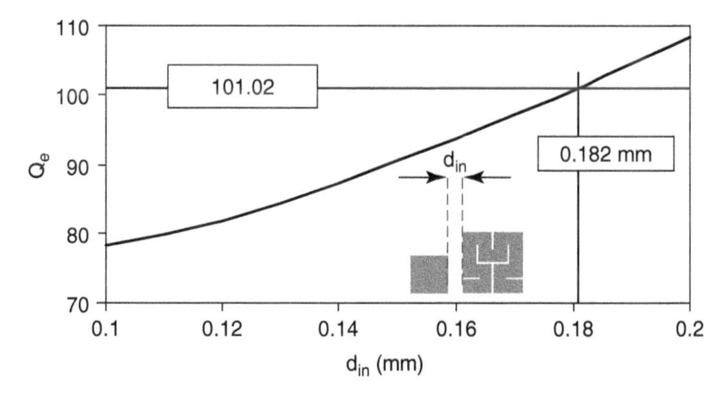

Figure 6.3 Matching the coupling between input line/first resonator and external quality factor.

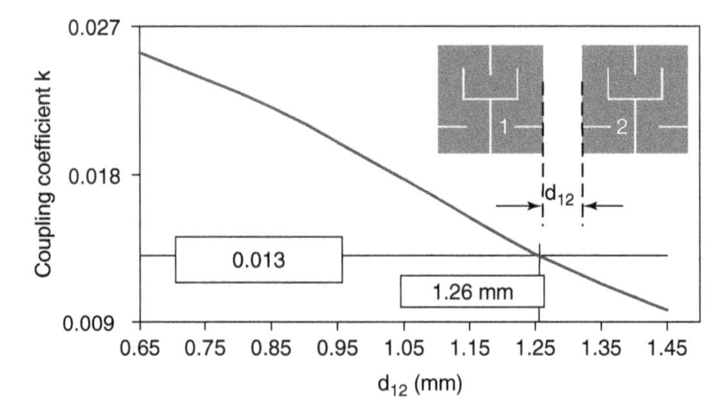

Figure 6.4 Matching the coupling between resonators.

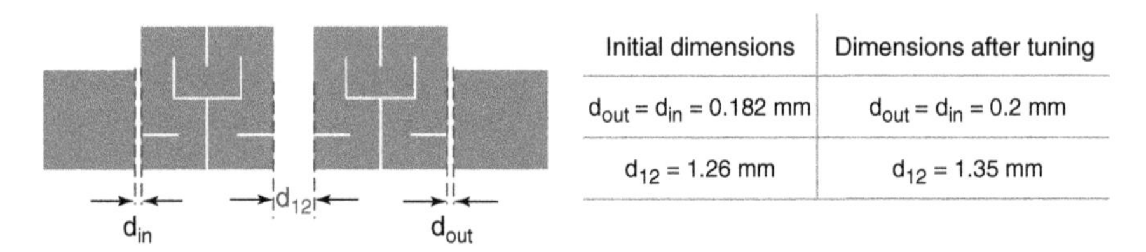

Initial dimensions	Dimensions after tuning
$d_{out} = d_{in} = 0.182$ mm	$d_{out} = d_{in} = 0.2$ mm
$d_{12} = 1.26$ mm	$d_{12} = 1.35$ mm

Figure 6.5 Second-order filter with gap adjustments.

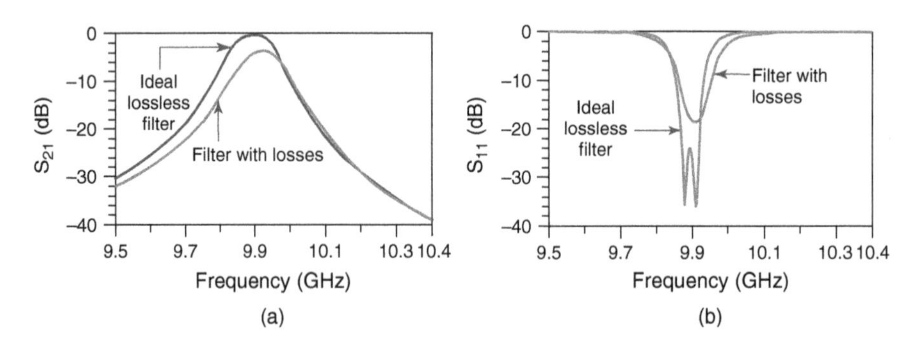

Figure 6.6 Simulated responses of the second-order Chebyshev filter.

When considering dielectric and metallic losses, the in-band insertion losses increase to 3.4 dB, and the reflection losses drop to −18.5 dB.

Figure 6.7 provides the wideband transmission characteristic of the filter. A transmission zero appears at 11.6 GHz and to about 18 GHz, the attenuation is better than 60 dB. There is a gap in the response around 20.2 GHz, the resonant frequency of the enclosure. Momentum software has problems computing the response around the resonant frequency of the enclosure.

To compare the Hilbert filter with a classical planar filter, a filter with the same desired filtering characteristics was designed using straight $\lambda/2$ line resonators.

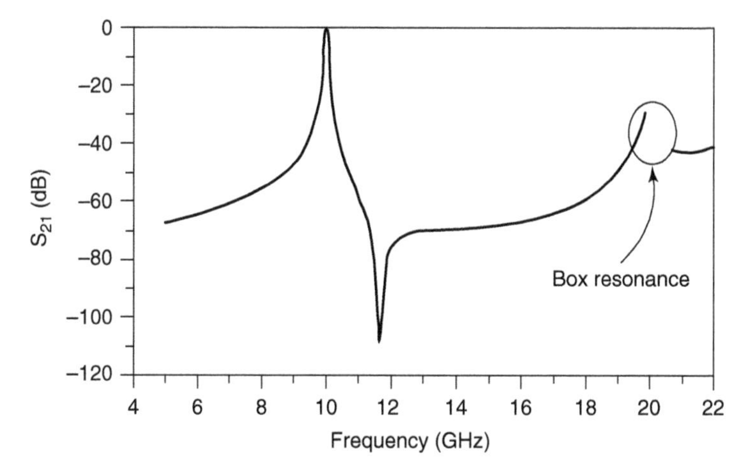

Figure 6.7 Wideband transmission characteristic of the second-order Hilbert filter.

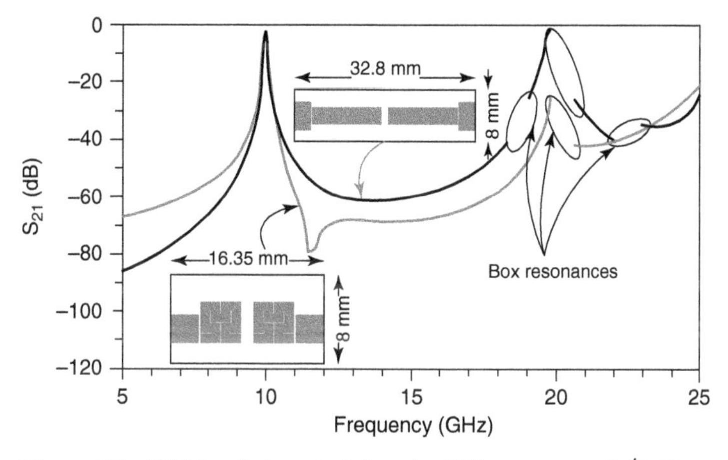

Figure 6.8 Wideband characteristics of a Hilbert versus a $\lambda/2$ filter.

The enclosure needed a length of 32.8 mm and a width of 8 mm to fit this filter. Figure 6.8 provides the wideband responses of the second-order Hilbert and $\lambda/2$ line resonator filters.

The insertion loss of the $\lambda/2$ line resonator filter is about 2 dB, compared to 3.1 dB for the Hilbert filter. This can be explained by the decrease in the unloaded quality factor when fractal iteration is applied. On the other hand, the Hilbert filter has better selectivity near the passband and higher attenuation in the stopband and a first spurious response that occurs at higher frequencies. One also can see that the enclosure resonance around 20 GHz is much more severe for the $\lambda/2$ line resonator filter. Except for the insertion loss, the Hilbert filter is better than the $\lambda/2$ line resonator filter, but the greatest advantage is that the Hilbert filter is half the size of the $\lambda/2$ line resonator filter.

6.2.2 Third-Order Chebyshev Responses

The Hilbert filter topology to provide a third-order Chebyshev response is shown in Figure 6.9. The filter is made of three aligned first-iteration Hilbert resonators and the input and output lines. The middle resonator is inverted compared to the other two. Figure 6.10 shows the bandpass prototype and the bandpass transmission and reflection responses for a third-order Chebyshev filter centered at 9.9 GHz with a 100-MHz bandwidth. The theoretical couplings needed for the design of the filter are $K_{12} = 0.01$ and $K_{23} = 0.01$. The theoretical external quality factor is $Q_e = 99$. The enclosure width is taken as 10 mm, and the length is estimated at 23.44 mm. In this case, enclosure resonance occurs at 15.8 and 19.1 GHz.

First, the distance d_{in} is defined by matching the coupling between the input line and the first resonator with the theoretical external quality factor $Q_e = 99$ using simulations as shown in Figure 6.11. This provides a distance d_{in} equal to 0.175 mm. Then the distance d_{12} is defined by matching the coupling between first and second resonators with the theoretical coupling coefficient $k_{12} = 0.01$ using simulations as shown in Figure 6.12. This provides a distance d_{12} equal to 0.185 mm.

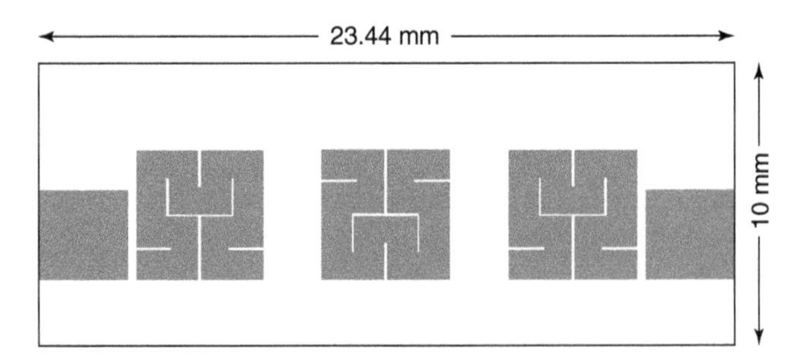

Figure 6.9 Hilbert filter topology for third-order Chebyshev responses.

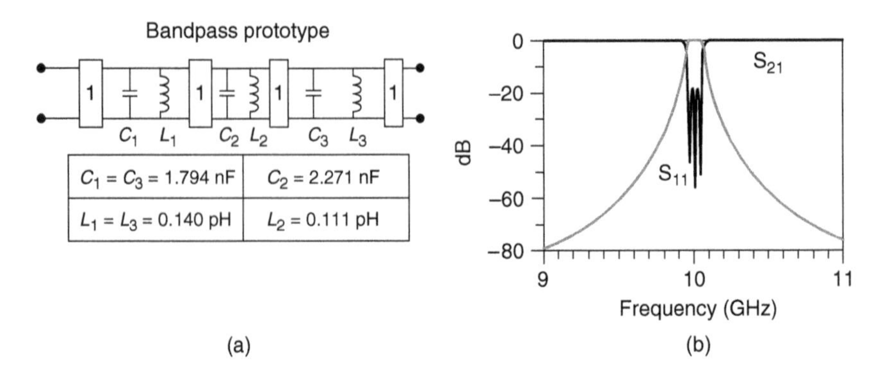

Figure 6.10 Third-order Chebyshev bandpass prototype.

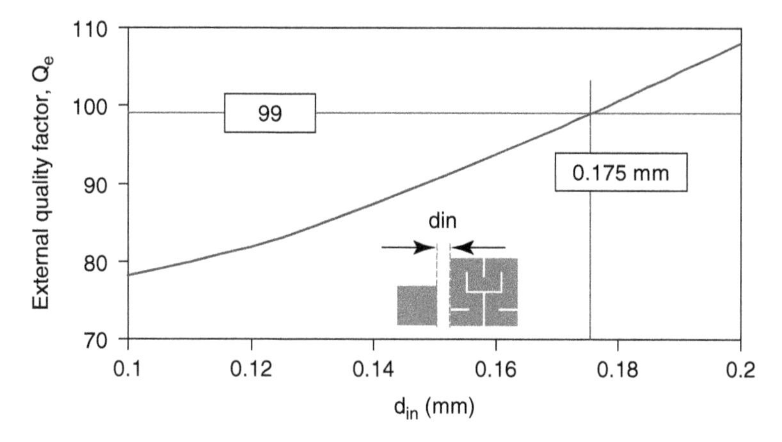

Figure 6.11 Matching input coupling and external quality factor.

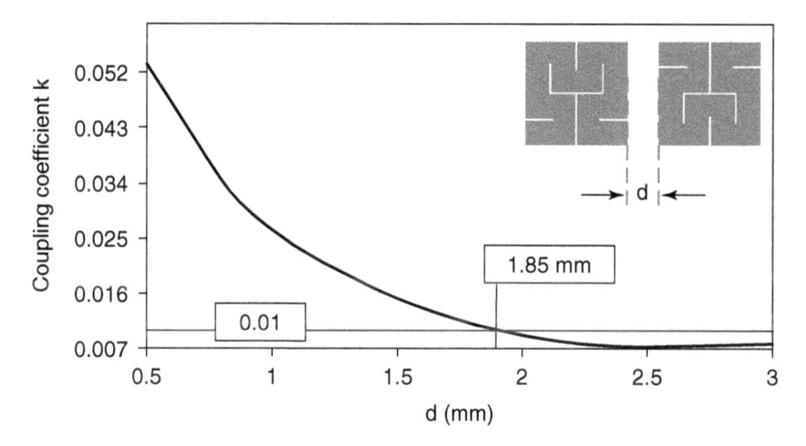

Figure 6.12 Matching the coupling between the first and second resonators to k_{12}.

The other gaps are obtained using the symmetry of the filter. As for the case of the second-order filter, when everything is simulated together, the actual couplings are stronger than expected and the gaps must be adjusted. Figure 6.13 provides the final dimensions of the gaps. Figure 6.14 shows the transmission and reflection characteristics of the filter both with and without dielectric and metallic losses. When considering dielectric and metallic losses, the in-band insertion loss increase to 5.8 dB and the reflection losses drop to −15 dB. The simulated wideband transmission characteristic is shown in Figure 6.15. The wideband characteristics show that a transmission zero appears at 5.7 GHz and that about −30 dB of attenuation is achieved at the resonant frequencies of the enclosure.

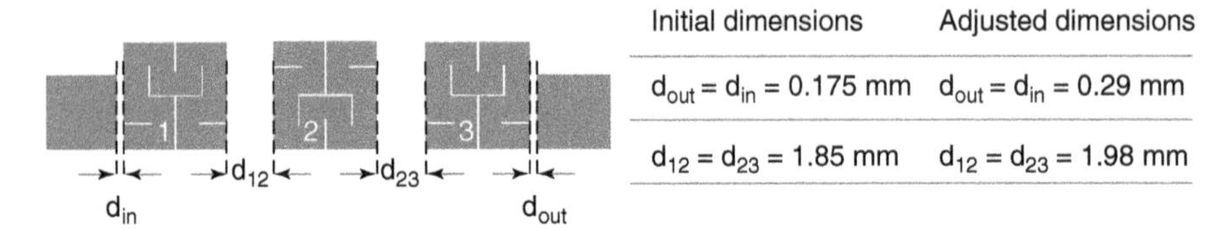

Initial dimensions	Adjusted dimensions
$d_{out} = d_{in} = 0.175$ mm	$d_{out} = d_{in} = 0.29$ mm
$d_{12} = d_{23} = 1.85$ mm	$d_{12} = d_{23} = 1.98$ mm

Figure 6.13 Third-order filter with gap adjustments.

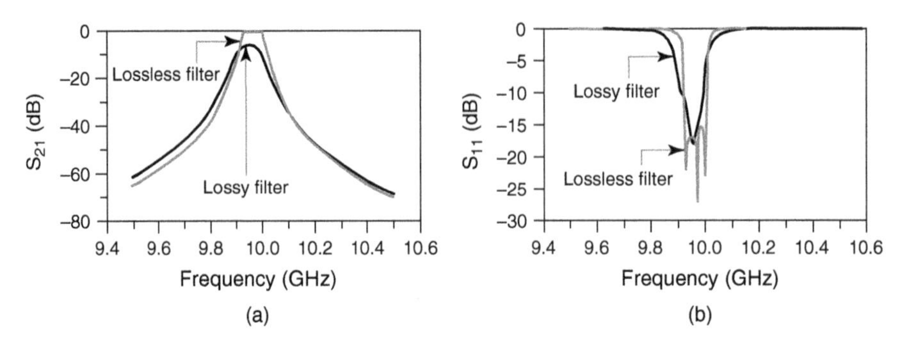

Figure 6.14 Simulated responses of the third-order Chebyshev filter.

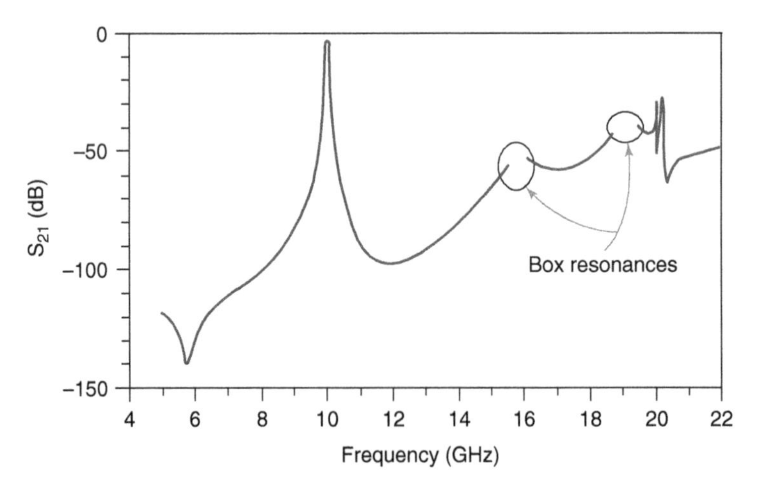

Figure 6.15 Wideband transmission characteristic of the third-order Hilbert filter.

6.2.3 Third-Order Pseudo-elliptic Responses

Zero on the Right Side of the Passband (Magnetic Coupling) The Hilbert filter topology to produce a third-order pseudo-elliptic response with the transmission zero on the right of the passband is shown in Figure 6.16. The cross-coupling between resonators 1 and 3 creates a transmission zero on the right side of the passband by using magnetic coupling between these resonators. Figure 6.17

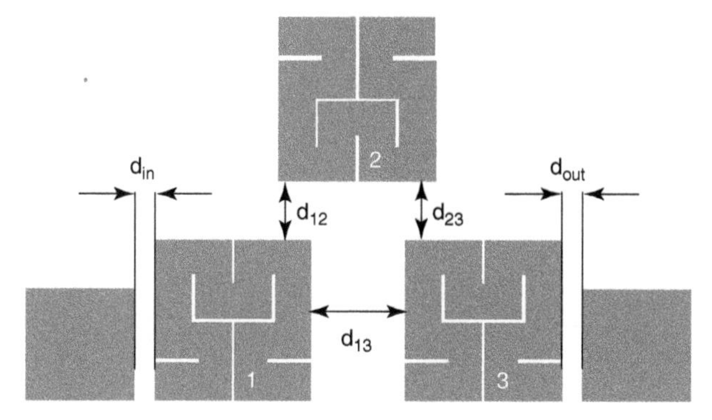

Figure 6.16 Hilbert filter topology for third-order pseudo-elliptic responses (magnetic coupling).

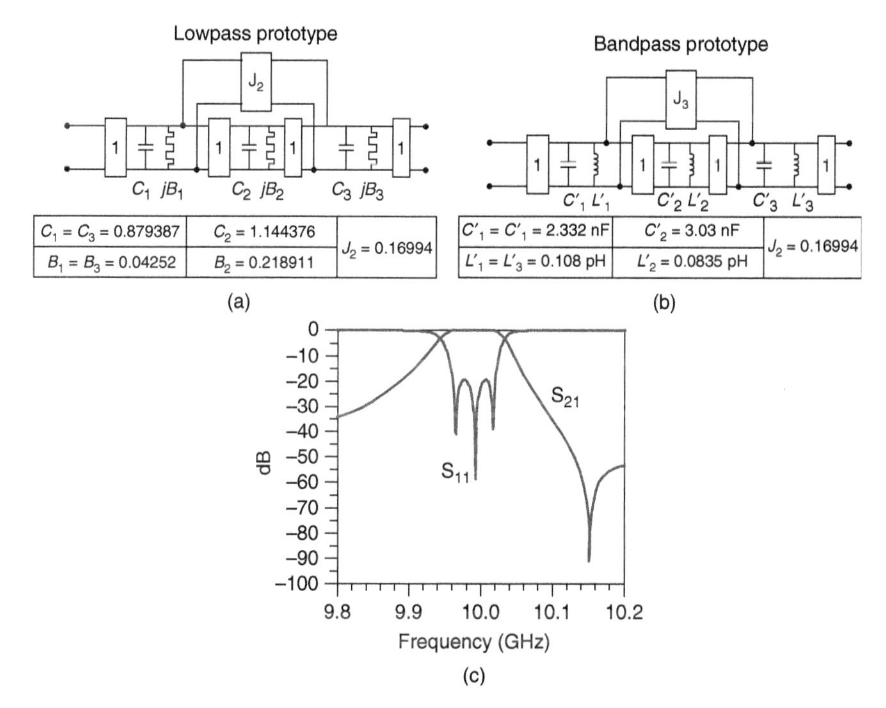

Figure 6.17 Third-order pseudo-elliptic prototype (zero on the right side).

shows the lowpass and bandpass prototypes for a third-order pseudo-elliptic filter centered at 10 GHz with a 100-MHz bandwidth and a transmission zero on the right side of the passband at 10.15 GHz. The bandpass transmission and reflection responses are also provided. The theoretical couplings needed for the design of the filter are $K_{12} = 0.0102$, $K_{13} = 0.0023$, and $K_{23} = 0.0102$. The theoretical external quality factor is $Q_e = 92.6$. The cross-coupling J_2 represents the coupling between resonators 1 and 3 and creates a transmission zero on the right side

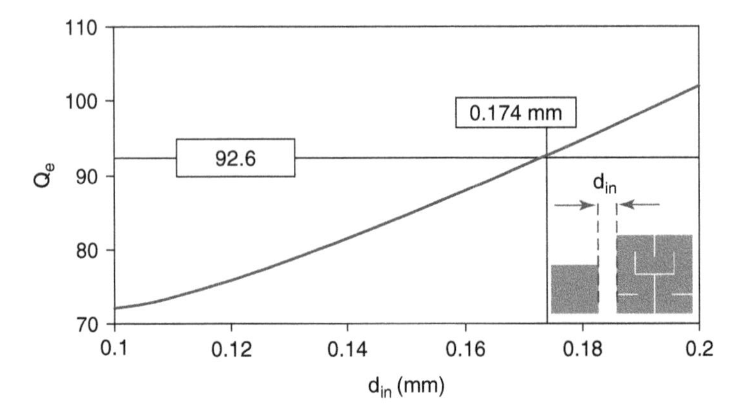

Figure 6.18 Matching input coupling and external quality factor.

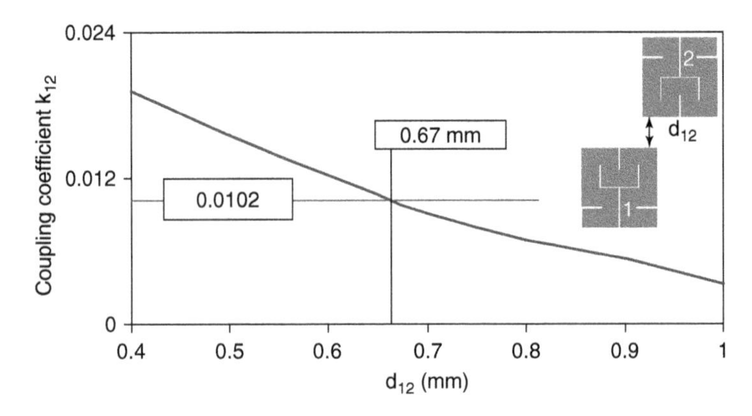

Figure 6.19 Matching the coupling between resonators 1 and 2 to k_{12}.

of the passband since it is positive. The enclosure width is taken as 14.64 mm and the length is estimated at 17.53 mm. The enclosure resonance occurs at 12.9 GHz.

First, the coupling between the input line and the first resonator is matched to the theoretical external quality factor $Q_e = 92.6$ using the simulations shown in Figure 6.18. This leads to a gap distance $d_{\text{in}} = 0.174$ mm. Next, the coupling between resonators 1 and 2 is matched to the theoretical coupling $k_{12} = 0.0102$ using the simulations shown in Figure 6.19. In this case we obtain a gap distance $d_{12} = 0.67$ mm. Finally, the coupling between resonators 1 and 3 is matched to the theoretical coupling $k_{13} = 0.0023$ using the simulations shown in Figure 6.20. In this case we obtain a gap distance $d_{13} = 2.45$ mm.

The other gaps are obtained by symmetry of the filter. Clearly, the simulations in Figures 6.18 to 6.20 occur in isolation from the other elements of the filter, and the gap distances must be adjusted using simulations of the entire filter. Figure 6.21 provides the final dimensions of the gaps. Figure 6.22 shows the

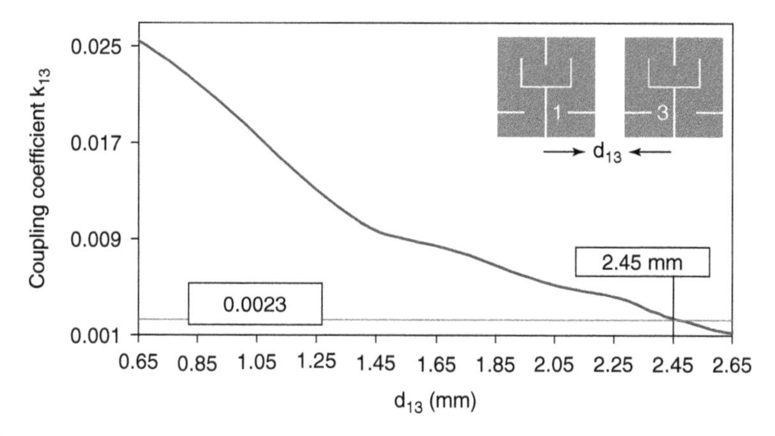

Figure 6.20 Matching the coupling between the first and third resonators to k_{13}.

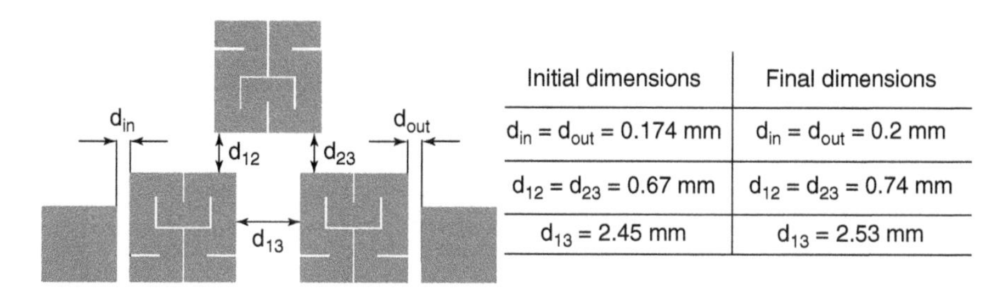

Initial dimensions	Final dimensions
$d_{in} = d_{out} = 0.174$ mm	$d_{in} = d_{out} = 0.2$ mm
$d_{12} = d_{23} = 0.67$ mm	$d_{12} = d_{23} = 0.74$ mm
$d_{13} = 2.45$ mm	$d_{13} = 2.53$ mm

Figure 6.21 Third-order filter with gap adjustments.

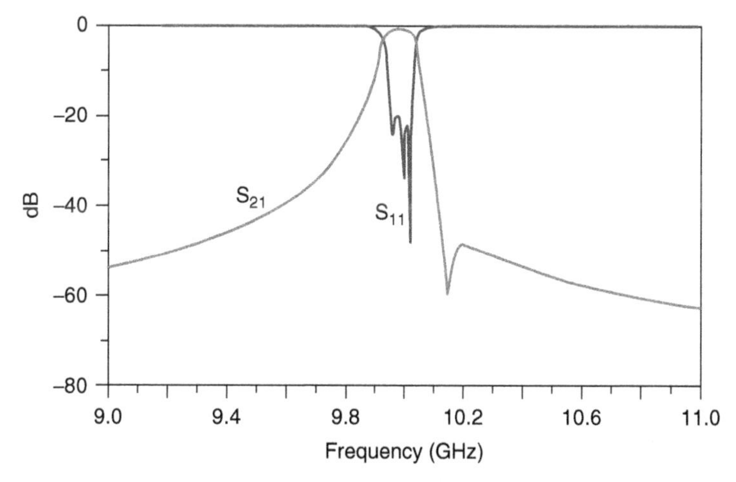

Figure 6.22 Simulated responses of the third-order pseudo-elliptic filter (zero on the right side).

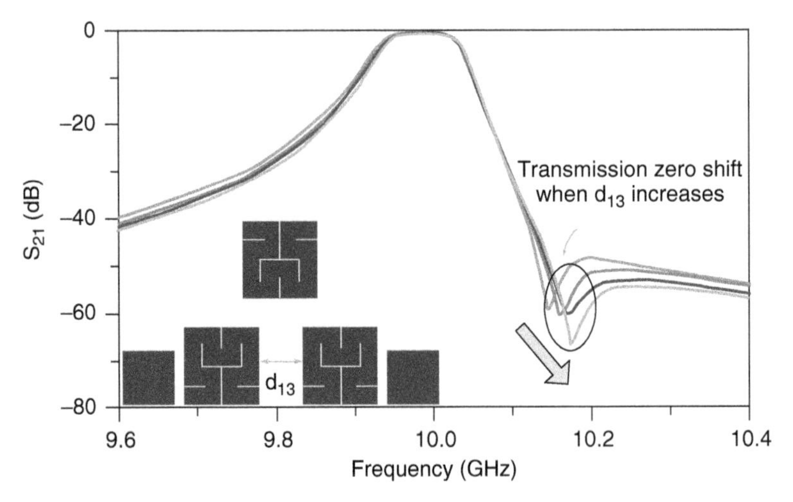

Figure 6.23 Effect of d_{13} on the location of the transmission zero.

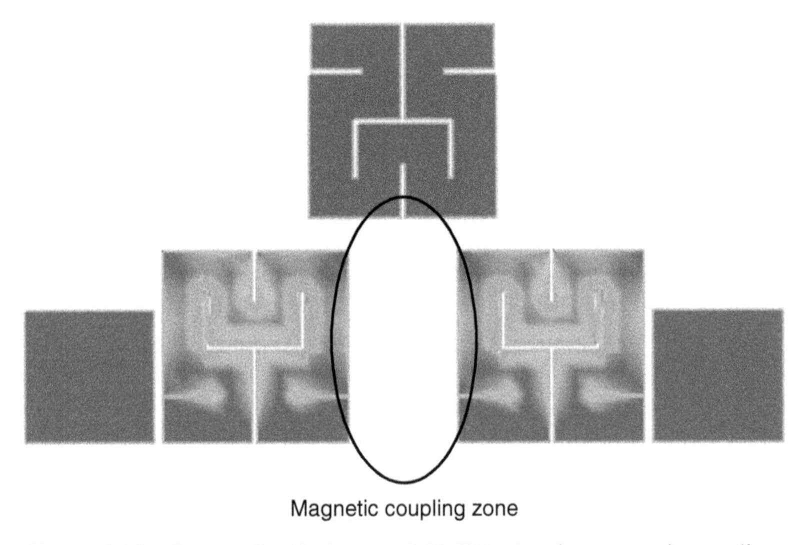

Figure 6.24 Current distribution at 10.12 GHz showing magnetic coupling.

simulated transmission and reflection characteristics of the filter. It is seen that the transmission zero is placed correctly at 10.15 GHz.

Figure 6.23 shows how d_{13} (coupling between the first and third resonators) controls well the location of the transmission zero without major changes to the rest of the response. The current distribution in the structure at 10.15 GHz is shown in Figure 6.24. These simulations show that the current is essentially concentrated on the outside of the coupling zone between resonators 1 and 3. Consequently, this coupling is considered magnetic.

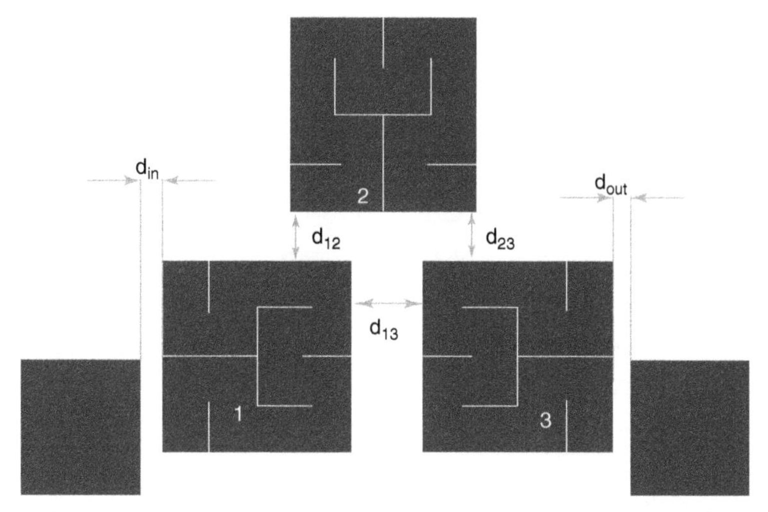

Figure 6.25 Hilbert filter topology for third-order pseudo-elliptic responses (electric coupling).

Zero on the Left Side of the Passband (Electric Coupling) The Hilbert filter topology to produce a third-order pseudo-elliptic response with the transmission zero on the left of the passband is shown in Figure 6.25. The cross-coupling between resonators 1 and 3 creates a transmission zero on the left side of the passband by using electric coupling between these resonators.

Figure 6.26 shows the lowpass and bandpass prototypes for a third-order pseudo-elliptic filter centered at 10 GHz with a 100-MHz bandwidth and with a transmission zero placed on the left side of the passband at 9.87 GHz. The bandpass transmission and reflection responses are also given. The theoretical couplings needed for the design of the filter are $K_{12} = 0.015$, $K_{13} = -0.006$, and $K_{23} = 0.015$. The theoretical external quality factor is $Q_e = 62$. The cross-coupling J_2 represents the coupling between resonators 1 and 3 and creates a transmission zero on the left side of the passband since it is negative. The enclosure width is taken as 15.1 mm and the length is estimated at 15.55 mm. The enclosure resonance occurs at 13.45 GHz.

The coupling between the input line and the first resonator is matched to the theoretical external quality factor $Q_e = 62$ using the simulations shown in Figure 6.27. This leads to a gap distance $d_{\text{in}} = 0.148$ mm. Next, the coupling between resonators 1 and 2 is matched to the theoretical coupling $k_{12} = 0.015$ using the simulations shown in Figure 6.28. In this case we obtain gap distance $d_{12} = 0.95$ mm. Finally, the coupling between resonators 1 and 3 is matched to the theoretical coupling $k_{13} = 0.006$ using the simulations shown in Figure 6.29. In this case we obtain gap distance $d_{13} = 1.04$ mm.

The other gaps are obtained by symmetry of the filter. Clearly, the simulations in Figures 6.27 to 6.29 occur in isolation from the other elements of the filter,

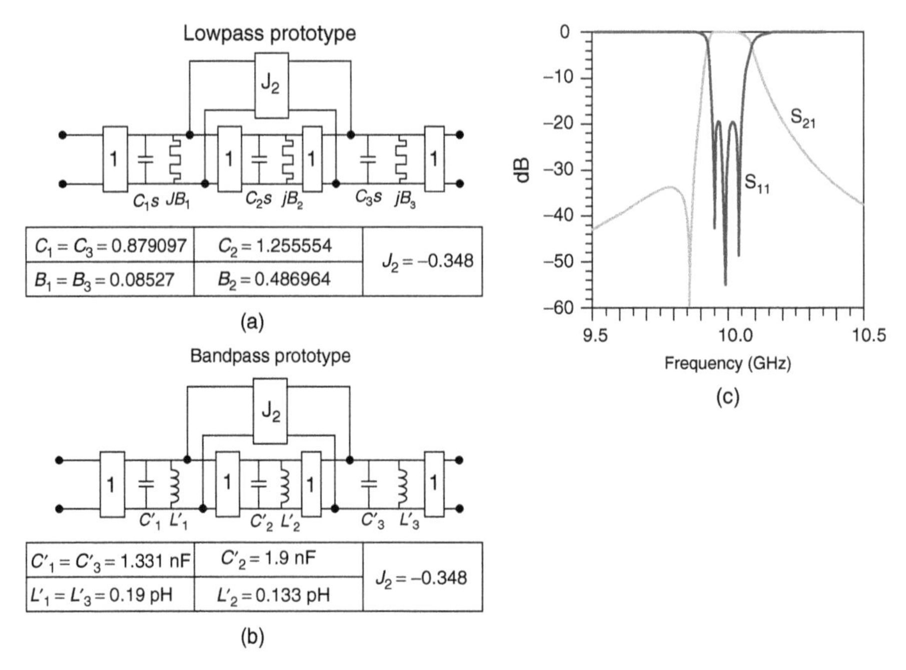

Figure 6.26 Third-order pseudo-elliptic prototype (zero on left side).

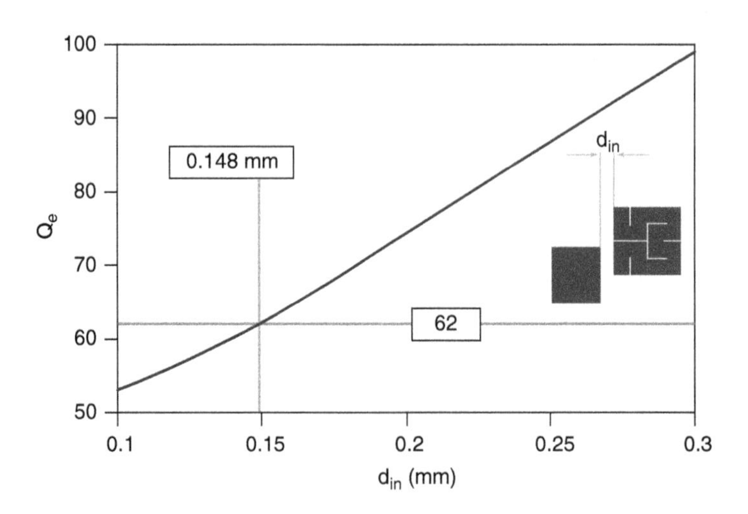

Figure 6.27 Matching input coupling and external quality factor.

and the gap distances must be adjusted using simulations of the entire filter. Figure 6.30 provides the final dimensions of the gaps. Figure 6.31 shows the simulated transmission and reflection characteristics of the filter. It is seen that the transmission zero is placed correctly at 9.87 GHz. The current distribution in the structure at 9.85 GHz is shown in Figure 6.32. These simulations show that

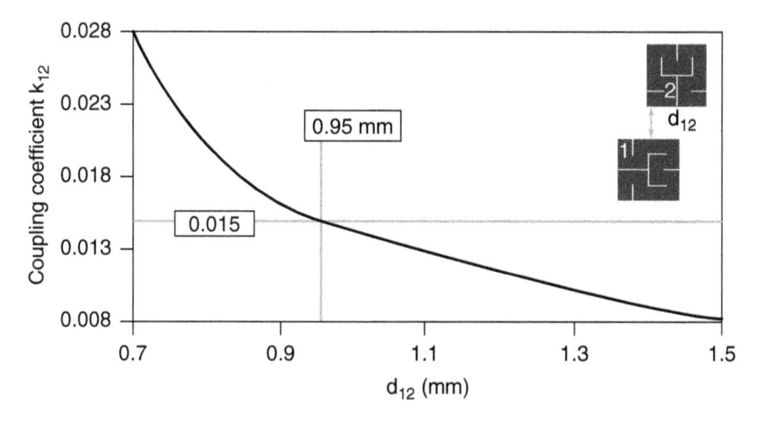

Figure 6.28 Matching the coupling between the first and second resonators to k_{12}.

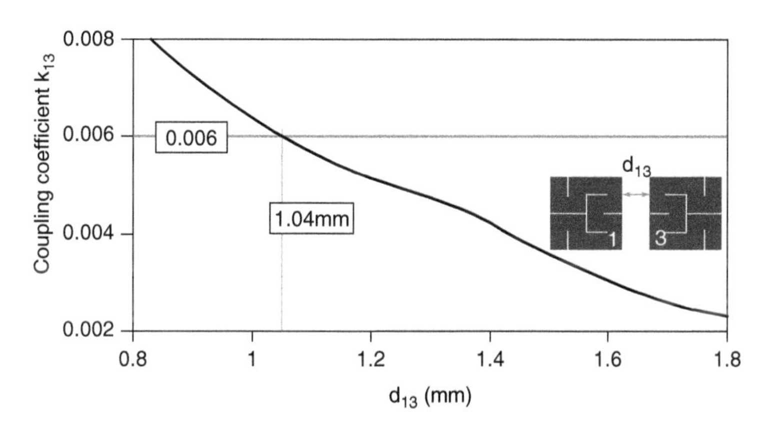

Figure 6.29 Matching the coupling between the first and third resonators to k_{13}.

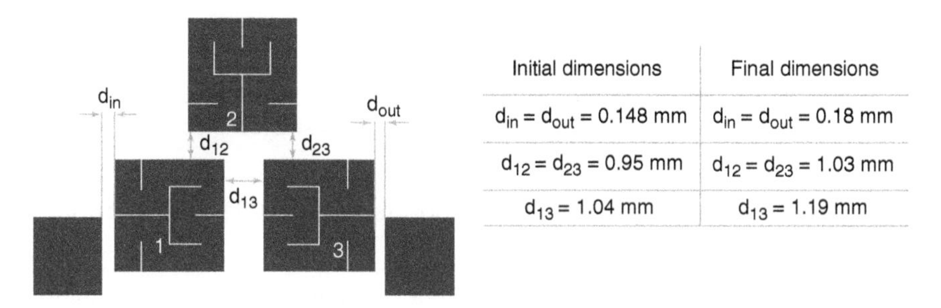

Initial dimensions	Final dimensions
$d_{in} = d_{out} = 0.148$ mm	$d_{in} = d_{out} = 0.18$ mm
$d_{12} = d_{23} = 0.95$ mm	$d_{12} = d_{23} = 1.03$ mm
$d_{13} = 1.04$ mm	$d_{13} = 1.19$ mm

Figure 6.30 Third-order filter with gap adjustments.

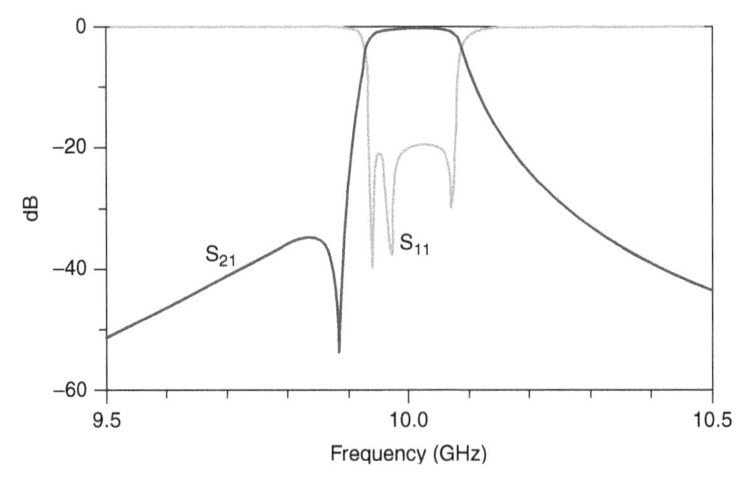

Figure 6.31 Simulated responses of the third-order pseudo-elliptic filter (zero on the left side).

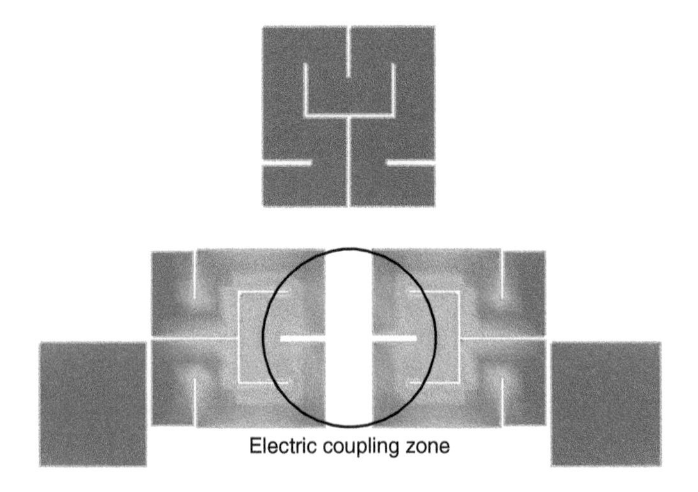

Figure 6.32 Current distribution at 9.85 GHz showing electric coupling.

the current is essentially concentrated on the inside of the coupling zone between resonators 1 and 3. Consequently, this coupling is considered electric.

6.2.4 Third-Order Pseudo-elliptic Responses Using Second-Iteration Resonators

The Hilbert filter topology for pseudo-elliptic responses using second-iteration resonators is shown in Figure 6.33. The coupling between resonators 1 and 3 creates a transmission zero on the right side of the passband. Figure 6.34 shows

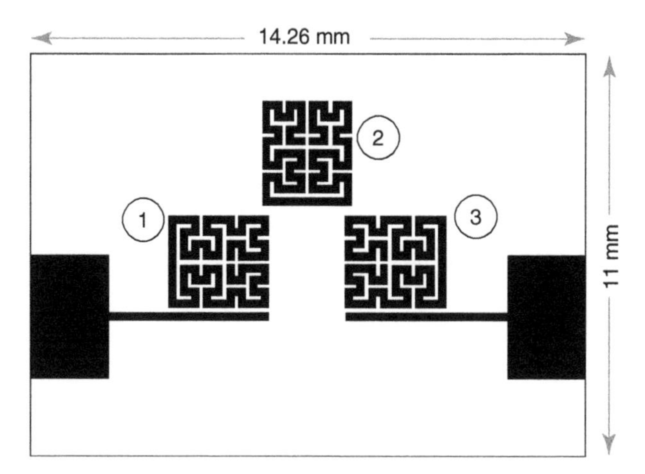

Figure 6.33 Hilbert filter topology for third-order pseudo-elliptic responses (magnetic coupling).

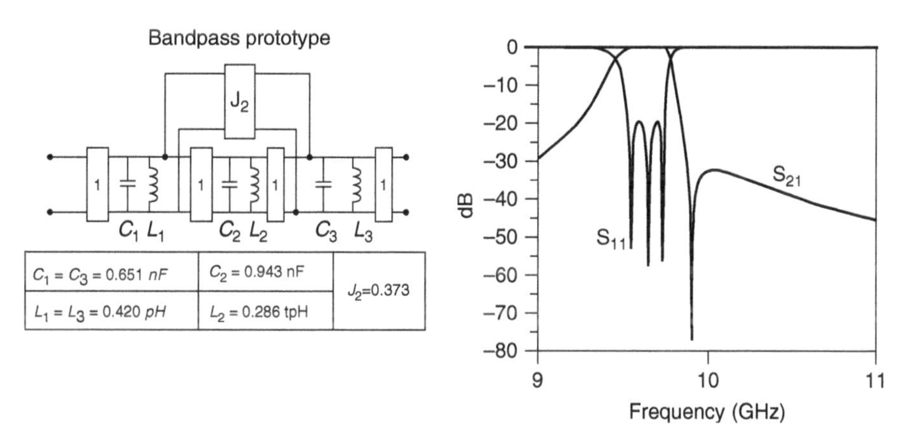

Figure 6.34 Third-order pseudo-elliptic bandpass prototype (zero on the right side).

the bandpass prototype for a third-order pseudo-elliptic filter centered at 9.63 GHz with a 200-MHz bandwidth. The transmission zero is placed on the right of the passband at 9.9 GHz. It also shows the expected transmission and reflection characteristics of the prototype filter. The theoretical couplings needed for the design of the filter are $K_{12} = 0.021$, $K_{13} = 0.0053$, and $K_{23} = 0.021$. The theoretical external quality factor is $Q_e = 44.58$. The cross-coupling J_2 represents the coupling between resonators 1 and 3 and creates a transmission zero on the right side of the passband since it is positive.

The enclosure width is taken as 11 mm and the length is estimated at 14.26 mm. The coupling between the input line and the first resonator is matched to the theoretical external quality factor $Q_e = 44.48$ using the simulations shown in Figure 6.35. This leads to a gap distance $d_{\text{in}} = 0.117$ mm. The

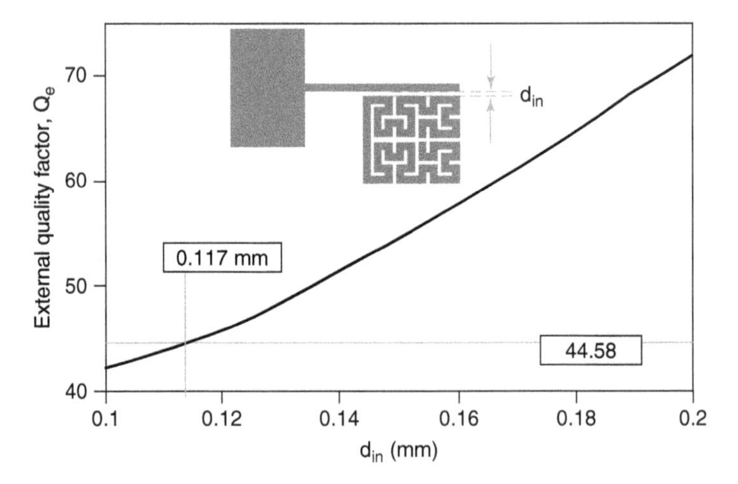

Figure 6.35 Matching input coupling and external quality factor.

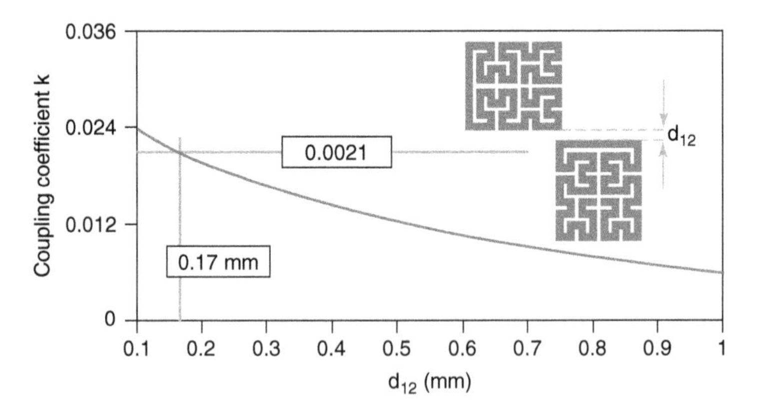

Figure 6.36 Matching the coupling between the first and second resonators to k_{12}.

coupling between resonators 1 and 2 is matched to the theoretical coupling $k_{12} = 0.021$ using the simulations shown in Figure 6.36. In this case, we obtain $d_{12} = 0.17$ mm. The coupling between resonators 1 and 3 is matched to the theoretical coupling $k_{13} = 0.0053$ using the simulations shown in Figure 6.37. In this case, we obtain $d_{13} = 1.69$ mm.

Since the simulations of Figures 6.35 to 6.37 were obtained in isolation from the other elements of the filter, a final adjustment of the gap values is done using simulations of the entire structure. The final filter gap dimensions are given in Figure 6.38. Figure 6.39 shows the transmission and reflection characteristics of the filter with and without taking dielectric and metallic losses into account. When considering dielectric and metallic losses, the in-band insertion loss is 7.2 dB and the return loss drops to −14 dB. The relatively high losses are due to a weak unloaded quality factor for higher iteration resonators. The use of gold as

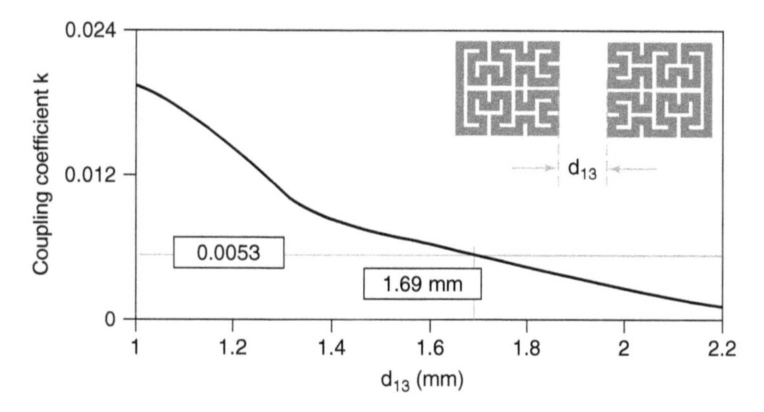

Figure 6.37 Matching the coupling between the first and third resonators to k_{13}.

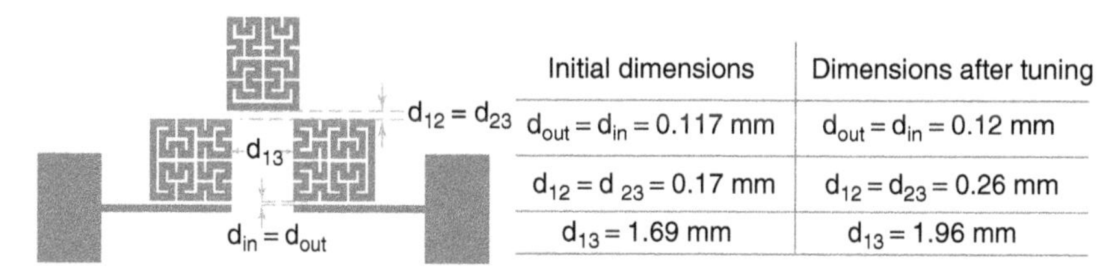

Initial dimensions	Dimensions after tuning
$d_{out} = d_{in} = 0.117$ mm	$d_{out} = d_{in} = 0.12$ mm
$d_{12} = d_{23} = 0.17$ mm	$d_{12} = d_{23} = 0.26$ mm
$d_{13} = 1.69$ mm	$d_{13} = 1.96$ mm

Figure 6.38 Third-order filter with gap adjustments.

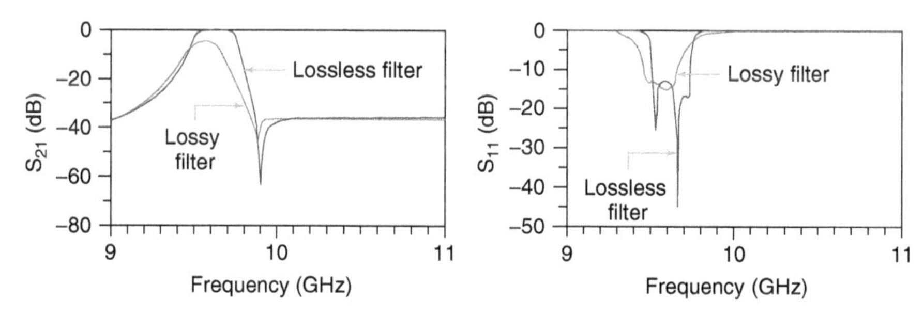

Figure 6.39 Simulated responses of the third-order pseudo-elliptic filter (zero on the right side).

a conducting metal and a thinner substrate with a low-loss tangent could improve these results.

The transmission zero can again be positioned accurately by adjusting the gap d_{13} as shown in the simulations of Figure 6.40. Simulations of the current distribution at the transmission zero of 9.9 GHz are shown in Figure 6.41. The current is essentially concentrated on the outside of the coupling zone between resonators 1 and 3, and is consistent with magnetic coupling. The wideband transmission characteristics are shown in Figure 6.42. The first spurious response

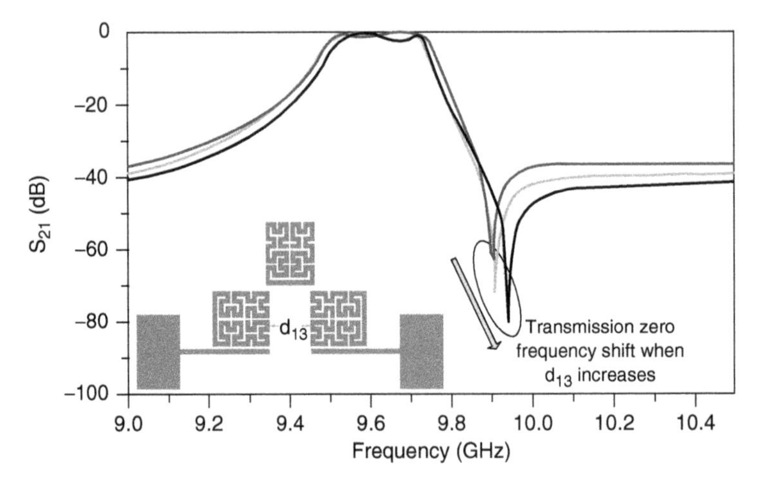

Figure 6.40 Effect of d_{13} on the location of the transmission zero.

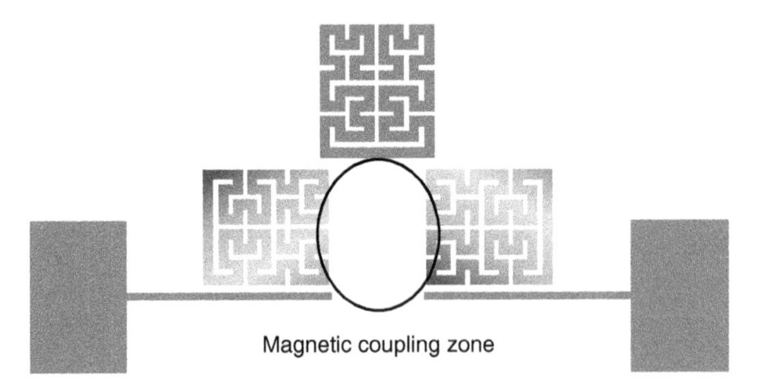

Figure 6.41 Current distribution at 9.9 GHz showing magnetic coupling.

appears at 20.5 GHz, with attenuation superior to 25 dB between 10 and 20 GHz, and superior to 35 dB between 5 and 9 GHz.

6.3 REALIZATIONS AND MEASURED PERFORMANCE

For cost reasons, copper was used for the suspended substrate stripline [1]. The filters were assembled in a laboratory environment [2–10], and great caution was needed not to twist the thin flexible substrate. The details of the fabrication process are given in Chapter 5.

The fabrication followed the same procedures as in Section 5.2.6. The enclosure was built in-house and the substrate and planar resonators (the filter) were fabricated by Atlantec.

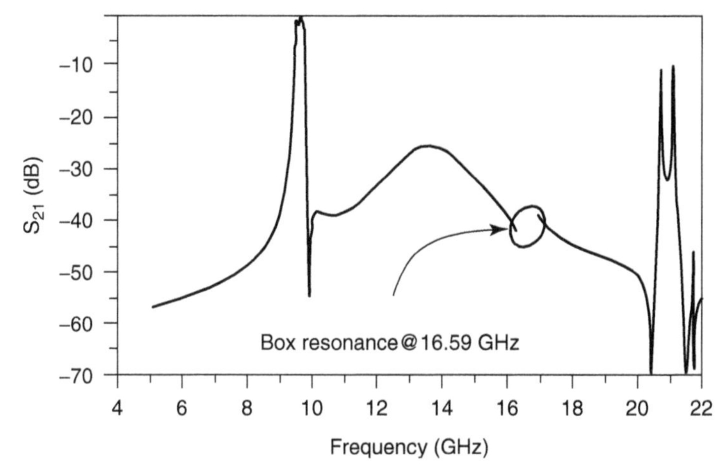

Figure 6.42 Wideband response of the third-order pseudo-elliptic filter (zero on the right).

6.3.1 Third-Order Pseudo-elliptic Filter with Transmission Zero on the Right

The example of the first filter presented in Section 6.2.3 was constructed and its performance was measured. The enclosure had a width of 14.64 mm and a length of 17.53 mm. Duroid substrate with a relative dielectric permittivity of 2.2 and a thickness of 0.254 mm was used. The upper and lower cavity heights were 1 mm. The gaps between resonators are given in Figure 6.43. The simulated response of this filter is shown in Figure 6.44, and the filter constructed is shown in Figure 6.45.

The measured and simulated wideband transmission characteristics are provided in Figure 6.46, and the measured and simulated reflection characteristics in the vicinity of the passband are given in Figure 6.47. The simulations in this case take into account dielectric and metallic losses. Table 6.1 summarizes the key features of the measured and simulated filters.

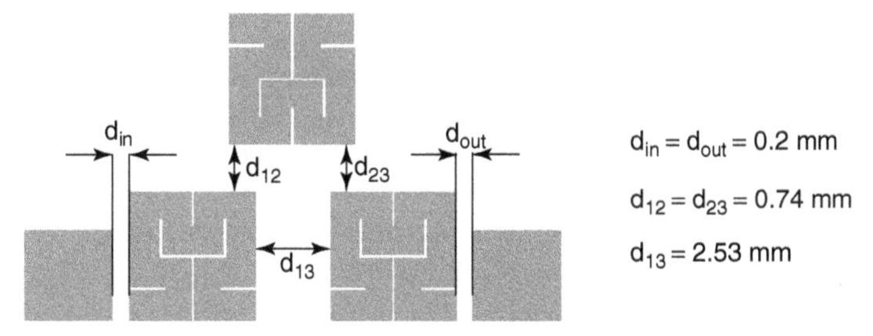

Figure 6.43 Third-order pseudo-elliptic Hilbert filter with a transmission zero on the right.

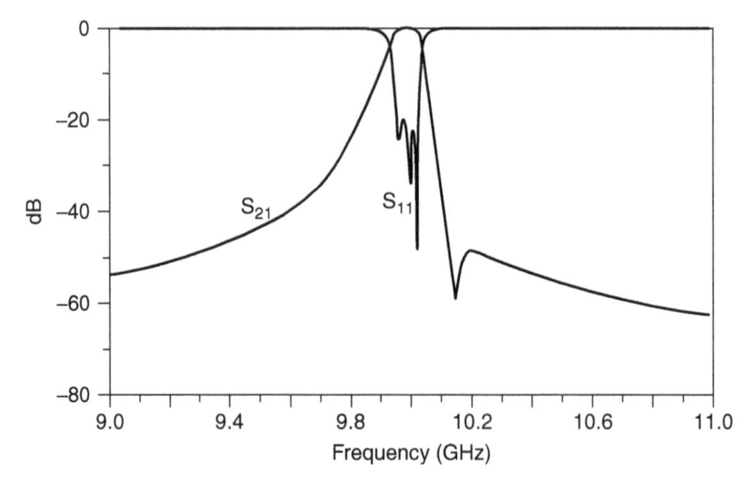

Figure 6.44 Simulated responses of the filter.

Figure 6.45 Realization of the third-order pseudo-elliptic Hilbert filter (zero on the right).

The measurements show that the central frequency and the transmission zero are higher than in the simulations. The measured passband is also larger than expected and suggests that the couplings are stronger than expected. The insertion loss measured is better than the loss expected. Since the simulations of Chapter 3 showed that for the same gap spacing, the coupling coefficient is stronger when the cavities are higher, we can think that the actual cavity is higher than expected. In addition, the unloaded quality factor of a resonator increases when the cavity heights are increased, thus reducing the insertion loss of the filter.

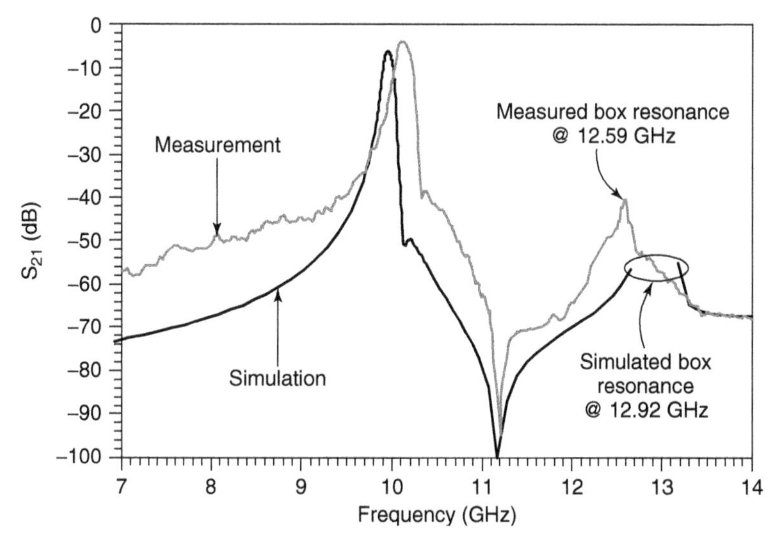

Figure 6.46 Measured and simulated wideband characteristics.

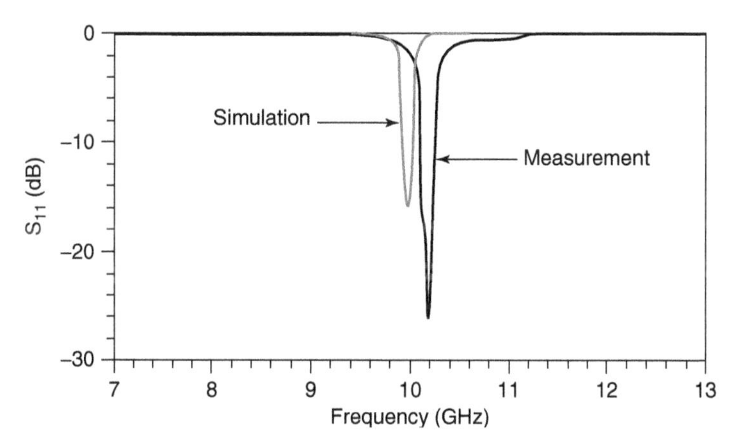

Figure 6.47 Measured and simulated return characteristics in the vicinity of the passband.

Finally, the measured enclosure resonance is lower than expected. This can be due to an enclosure with larger dimensions than expected. Next we look at a few sensitivity analyses to show where the differences between measured and simulations originated.

Figure 6.48 shows that when the cavity heights increase, the filter characteristics are shifted to higher frequencies and that bandwidth also increases, due to the increase in the coupling coefficients. The insertion loss decreases when the cavity heights increase since the unloaded quality factor of the resonators increases. Figure 6.49 shows that a reduction in

TABLE 6.1 Summary of Key Filter Characteristics

Filter Characteristic	Simulation	Measurement
Central frequency	10 GHz	10.12 GHz
Bandwidth	100 MHz	142 MHz
Insertion loss	−5.4 dB	−3.75 dB
Return loss	≤ −15 dB	≤ −17 dB
Transmission zero	10.145 GHz	10.33 GHz
Enclosure resonance	12.92 GHz	12.59 GHz

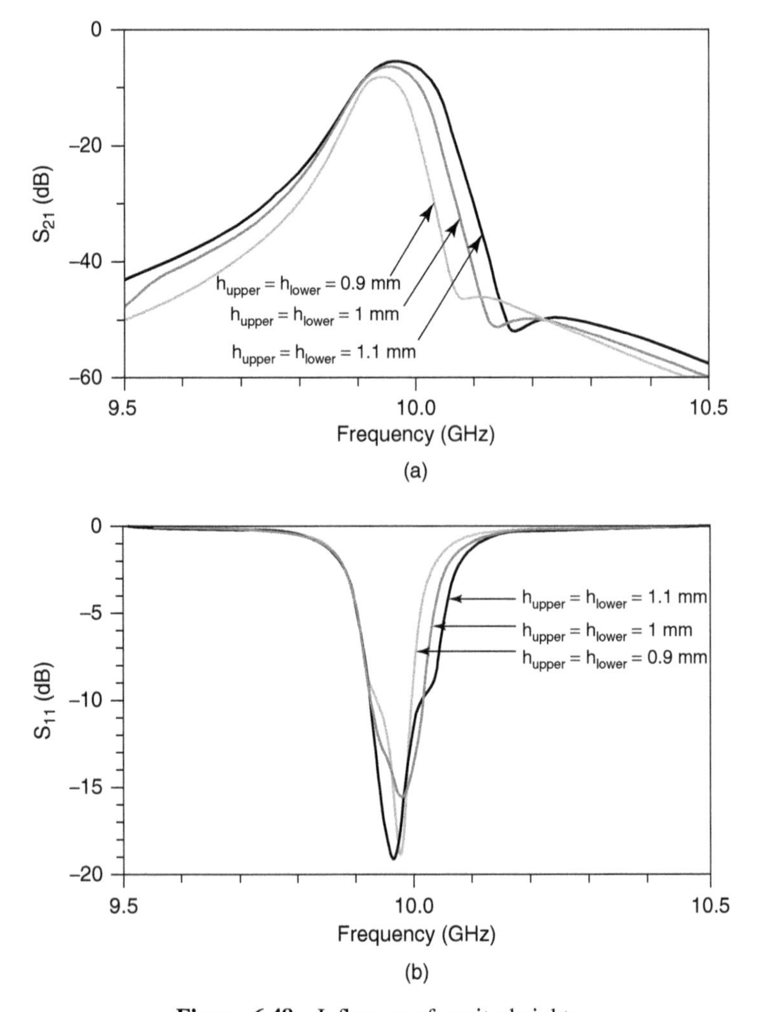

Figure 6.48 Influence of cavity heights.

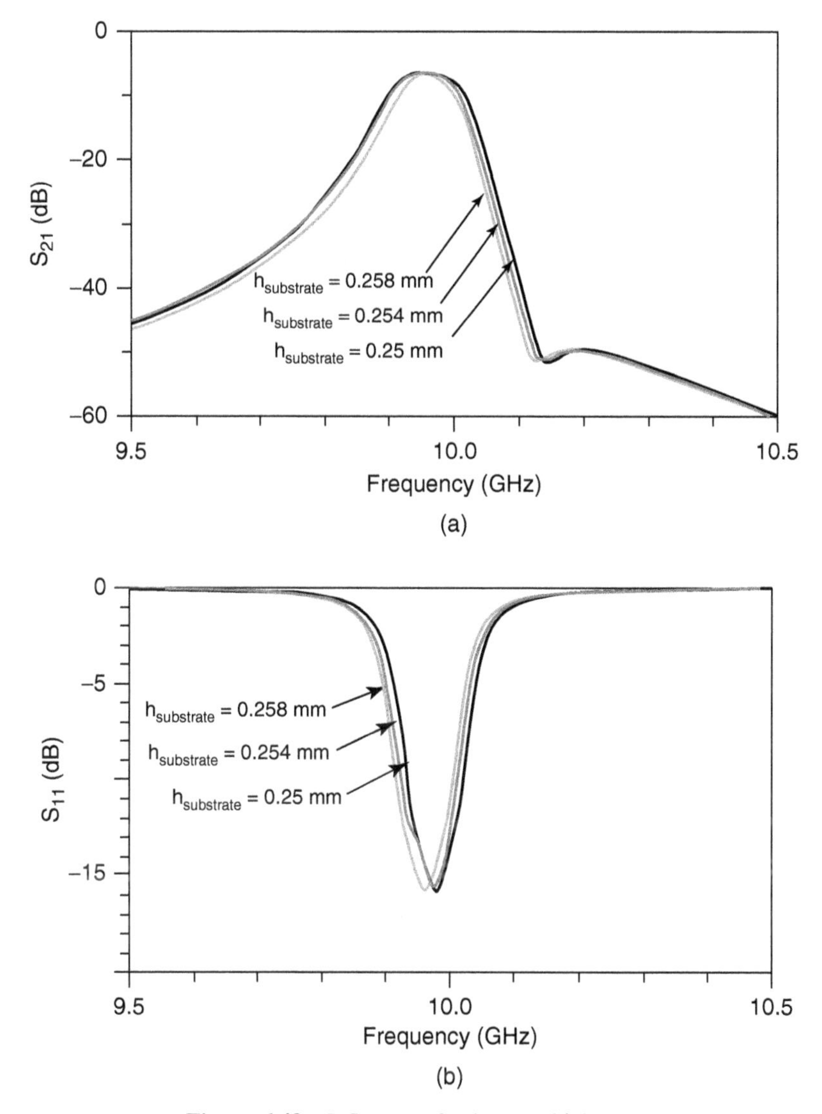

Figure 6.49 Influence of substrate thickness.

substrate thickness shifts the propagation characteristics of the filter at higher frequencies.

For the previous simulations, a mesh density of 35 cells/λ was used. Figure 6.50 shows simulations of the filter using different mesh densities. The propagation characteristics are shifted slightly toward higher frequencies when the mesh density increases. However, using a larger mesh density results in longer simulation times.

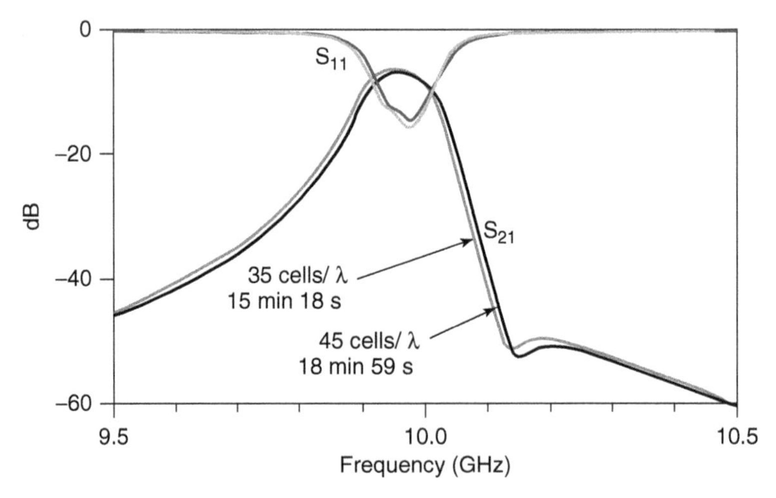

Figure 6.50 Influence of mesh density.

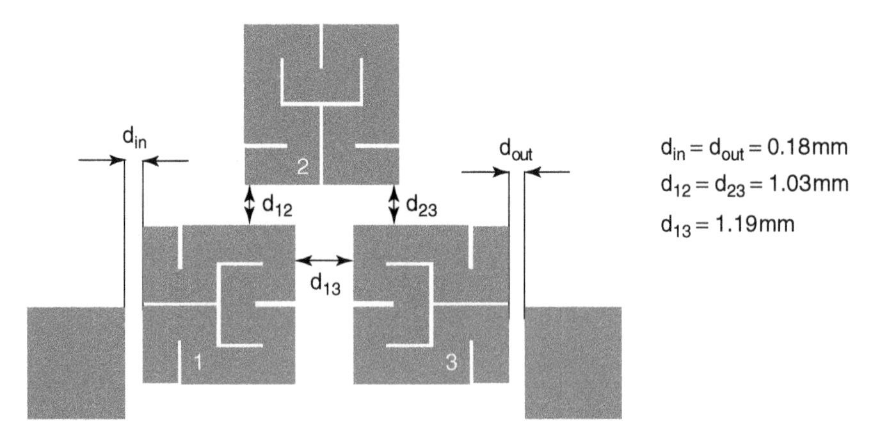

Figure 6.51 Third-order pseudo-elliptic Hilbert filter with transmission zero on the left.

6.3.2 Third-Order Pseudo-elliptic Filter with Transmission Zero on the Left

The example of the second filter presented in Section 6.2.3 was constructed and its performance was measured. The enclosure had a width of 15.1 mm and a length of 15.55 mm. Duroid substrate with a relative dielectric permittivity of 2.2 and a thickness of 0.254 mm was used. The upper and lower cavity heights were 1 mm. The gaps between resonators are given in Figure 6.51. The simulated

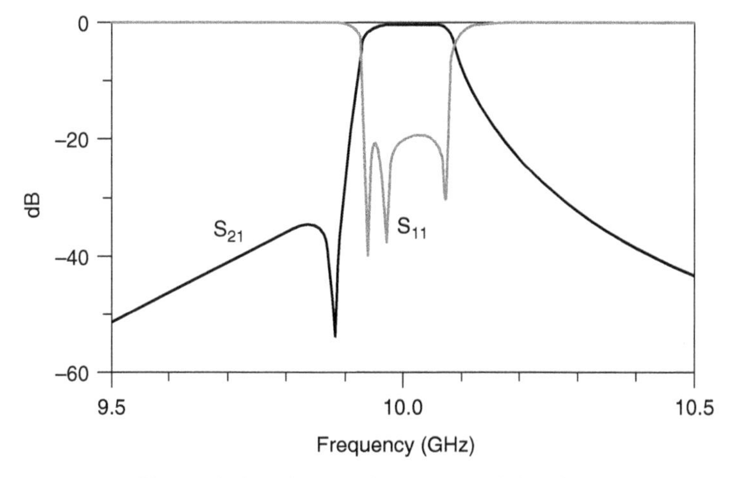

Figure 6.52 Simulated responses of the filter.

Figure 6.53 Realization of the third-order pseudo-elliptic Hilbert filter (zero on the left).

response of this filter is shown in Figure 6.52, and Figure 6.53 shows the filter realized.

Measurements were carried out using a HP8720D network analyzer. The measured and simulated transmission characteristics are provided in Figure 6.54, and the measured and simulated reflection characteristics are given in Figure 6.55.

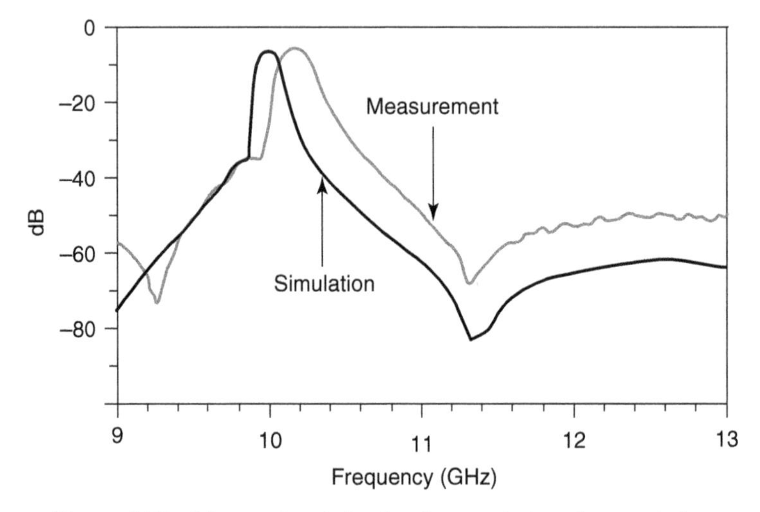

Figure 6.54 Measured and simulated transmission characteristics.

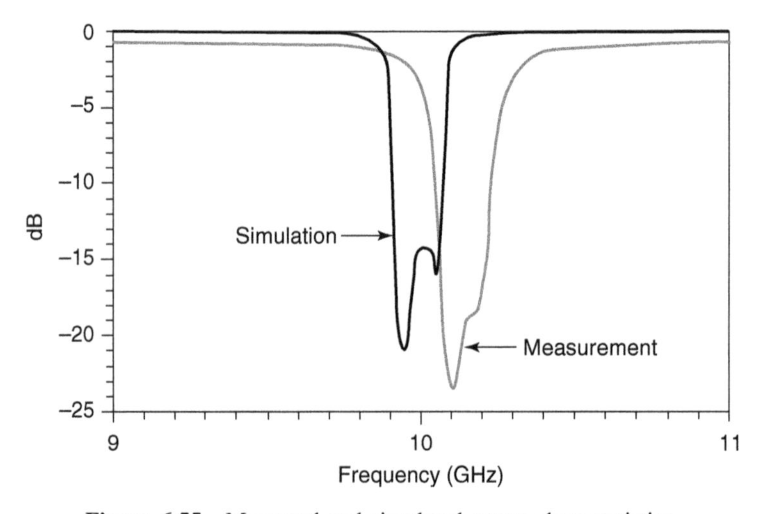

Figure 6.55 Measured and simulated return characteristics.

Figure 6.56 provides the measured and simulated wideband characteristics. The simulations in this case take into account dielectric and metallic losses. Table 6.2 summarizes the key features of the measured and simulated filters.

The measurements show that the central frequency and transmission zero are shifted toward higher frequencies. The measured passband is larger than the passband expected and the measured insertion loss is about as expected. The measured enclosure resonance is also lower than that expected. The differences between measured and simulated responses are quite similar to those observed

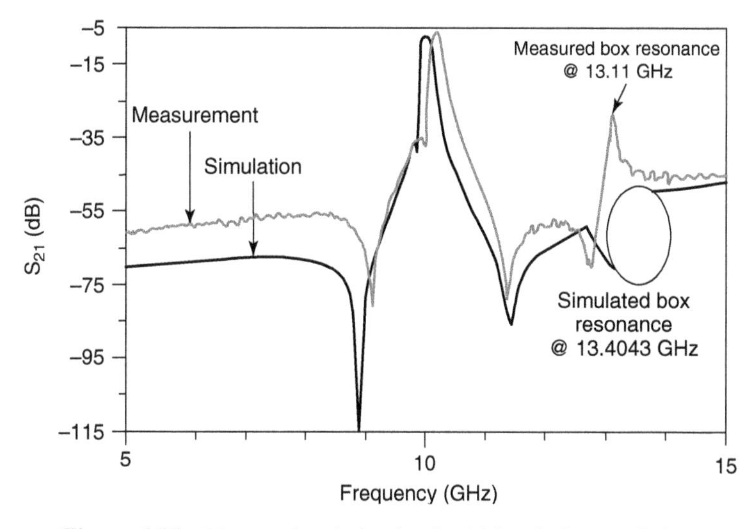

Figure 6.56 Measured and simulated wideband characteristics.

TABLE 6.2 Summary of Key Filter Characteristics

Filter Characteristic	Simulation	Measurement
Central frequency	10 GHz	10.18 GHz
Bandwidth	150 MHz	164 MHz
Insertion loss	−6.9 dB	−6.3 dB
Return loss	≤ −15 dB	≤ −19 dB
Transmission zero	9.87 GHz	9.99 GHz
Box resonance	13.4 GHz	13.11 GHz

for the previous filter, where larger cavity heights were thought to be the main cause of the problem.

Figure 6.57 shows that when the cavity heights increase, the filter characteristics are shifted to higher frequencies and that the bandwidth also increases, due to the increase in coupling coefficients. The insertion loss decreases when the cavity heights increase since the unloaded quality factor of the resonators increases. Figure 6.58 shows that a reduction in substrate thickness shifts the propagation characteristics of the filter at higher frequencies.

For the previous simulations, a mesh density of 35 cells/λ was used. Figure 6.59 shows simulations of the filter using different mesh densities. The propagation characteristics are shifted slightly toward higher frequencies when the mesh density increases. However, simulation time increases when increasing the mesh density.

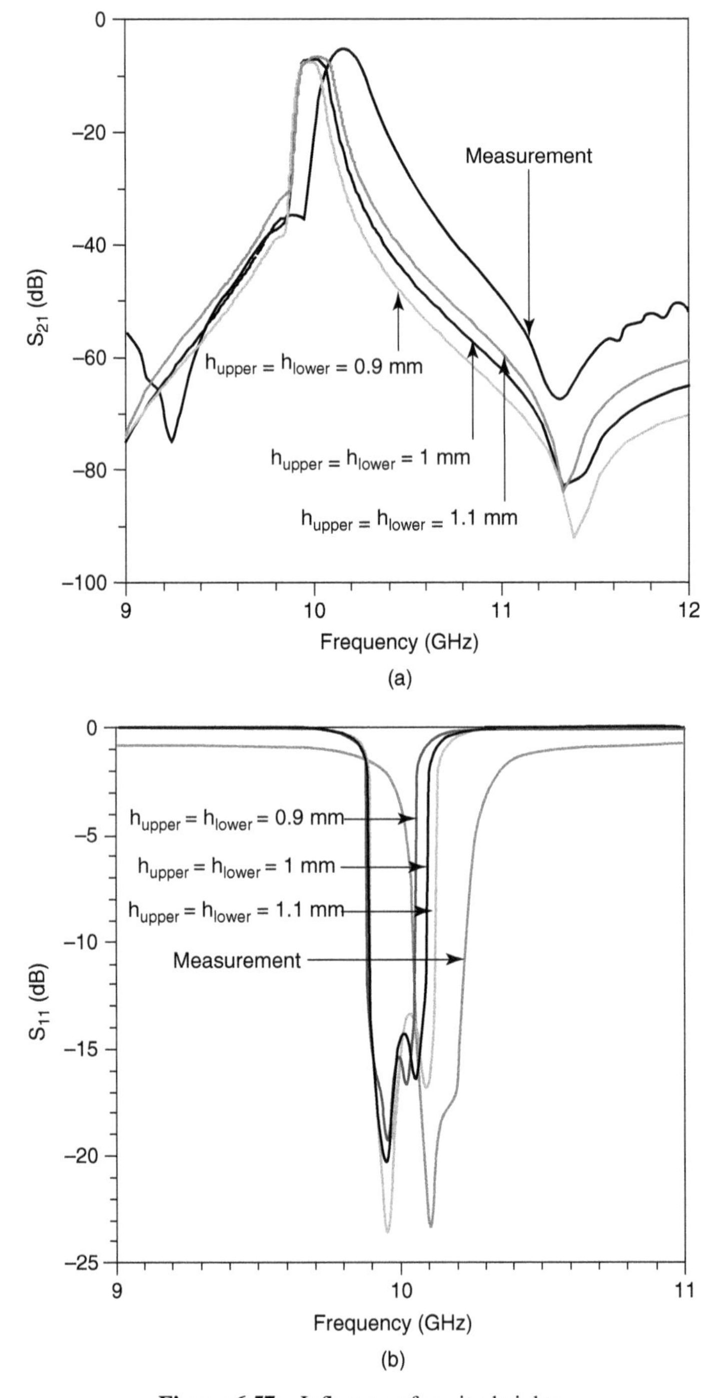

Figure 6.57 Influence of cavity heights.

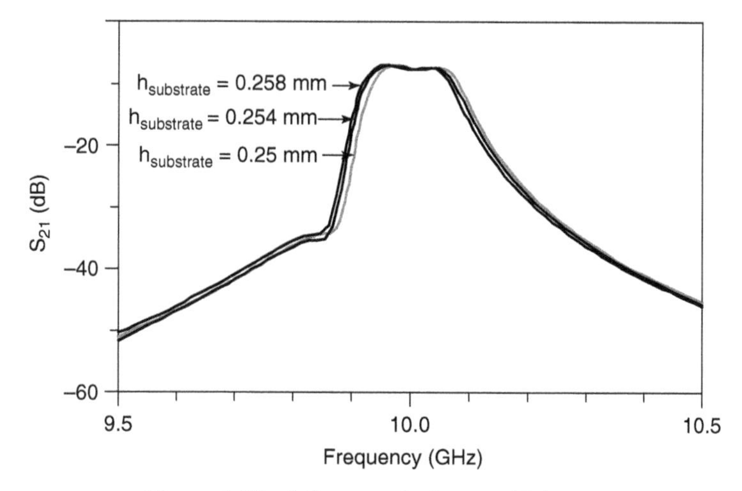

Figure 6.58 Influence of substrate thickness.

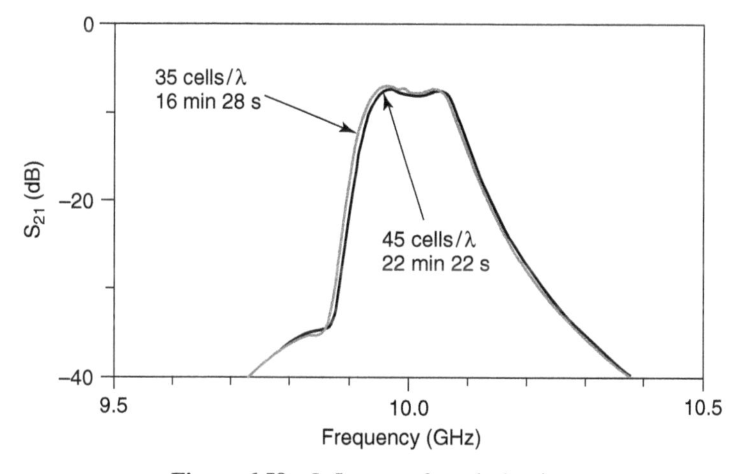

Figure 6.59 Influence of mesh density.

REFERENCES

[1] I. J. Bahl and P. Bhartia, *Microwave Solid State Circuit Design*, Wiley, Hoboken, NJ, 2004, Sec. 6.4.

[2] E. Hanna, P. Jarry, E. Kerhervé, J. M. Pham, and D. Lo Hine Tong, Synthesis and design of a suspended substrate capacitive-gap parallel coupled-line (CGCL) bandpass filter with one transmission zero, *10th IEEE International Conference on Electronics Circuits and Systems (ICECS2003)*, University of Sharjah, United Arab Emirates, pp. 547–550, Dec. 14–17, 2003.

[3] E. Hanna, P. Jarry, E. Kerhervé, and J. M. Pham, General prototype network method for suspended substrate microwave filters with asymmetrical prescribed transmission

zeros: synthesis and realization, *Mediterranean Microwave Symposium* (*MMS 2004*), Marseille, France, pp. 40–43, June 1–3, 2004.

[4] E. Hanna, P. Jarry, E. Kerhervé, and J. M. Pham, General prototype network with cross coupling synthesis method for microwave filters: application to suspended substrate strip line filters, *ESA–CNES International Workshop on Microwave Filters*, Toulouse, France, Sept. 13–15, 2004.

[5] E. Hanna, P. Jarry, E. Kerhervé, and J. M. Pham, Etude de résonateurs planaires fractals et application à la synthèse de filtres microondes en substrat suspendu, *14th National Microwave Days* (*JNM 2005*), Nantes, France, May 11–13, 2005, No. 7A4, Abstract, p. 134.

[6] E. Hanna, P. Jarry, E. Kerhervé, and J. M. Pham, Suspended substrate stripline cross-coupled trisection filters with asymmetrical frequency characteristics, *35th European Microwave Conference* (*EuMC*), Paris, pp. 273–276, June 4–6 2005.

[7] E. Hanna, P. Jarry, E. Kerhervé, and J. M. Pham, Cross-coupled suspended stripline trisection bandpass filters with open-loop resonators, *International Microwave and Optoelectronics Conference* (*IMOC 2005*), Brasília, Brazil, pp. 42–46, July 2005.

[8] E. Hanna, P. Jarry, E. Kerhervé, and J. M. Pham, Novel compact suspended substrate stripline trisection filters in the Ku band, *2005 Mediterranean Microwave Symposium* (*MMS 2005*), Athens, Greece, pp. 146–149, Sept. 6–8, 2005.

[9] E. Hanna, P. Jarry, E. Kerhervé, and J. M. Pham, A design approach for capacitive gap parallel-coupled line filters and fractal filters with the suspended substrate technology, in *Microwave Filters and Amplifiers* (scientific editor P. Jarry), Research Signpost, Tivandrum, India, pp. 49–71, 2005.

[10] J. M. Pham, P. Jarry, E. Kerhervé, and E. Hanna, A novel compact suspended stripline band pass filter using open-loop resonators technology with transmission zeroes, *International Microwave and Optoelectronics Conference* (*IMOC 2007*), Salvador, Brazil, pp. 937–940, Oct. 2007.

CHAPTER 7

DESIGN AND REALIZATION OF DUAL-MODE MINKOWSKI FILTERS

7.1 Introduction 161
7.2 Study of Minkowski dual-mode resonators 162
7.3 Design of fourth-order pseudo-elliptic filters with two transmission zeros 166
7.3.1 Topology 166
7.3.2 Synthesis 167
7.3.3 Design steps 168
7.4 Realization and measured performance 176
References 182

7.1 INTRODUCTION

In this chapter we study further the properties of the Minkowski resonators. We will see that by changing the symmetry of the resonator, two orthogonal modes of propagation can be decoupled. The use of dual-mode resonators allows additional reduction in the size of the filters since in general for an nth-order filter, only $n/2$ dual-mode resonators are necessary. In addition, these resonators can usually handle high power levels [1].

Design and Realizations of Miniaturized Fractal RF and Microwave Filters,
By Pierre Jarry and Jacques Beneat

7.2 STUDY OF MINKOWSKI DUAL-MODE RESONATORS

We first study the behavior of a dual-mode filter made of a single dual-mode resonator. Figure 7.1 shows the topology and frequency responses of several single dual-mode resonator filters. The first filter uses a dual-mode square resonator, the second uses a first-iteration dual-mode Minkowski resonator, and the third uses a second-iteration dual-mode Minkowski resonator. The top left corner of each resonator has been modified to introduce a dissymmetry in the

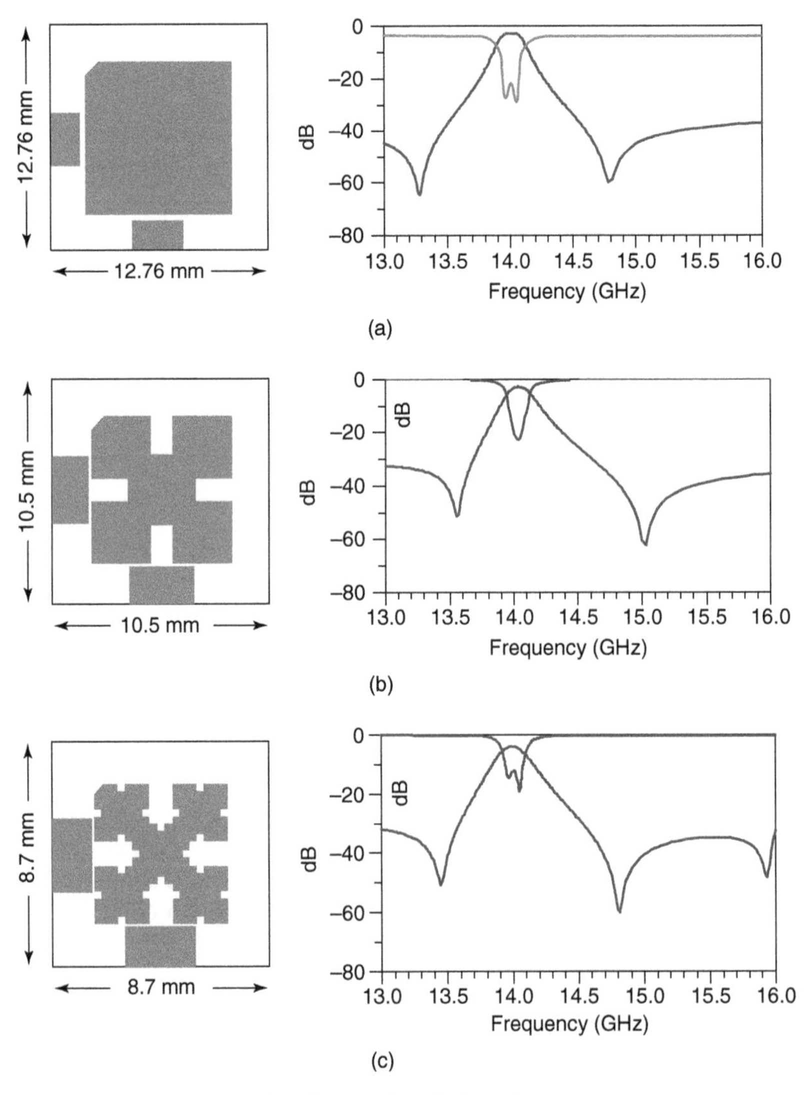

Figure 7.1 Filters using dual-mode resonators.

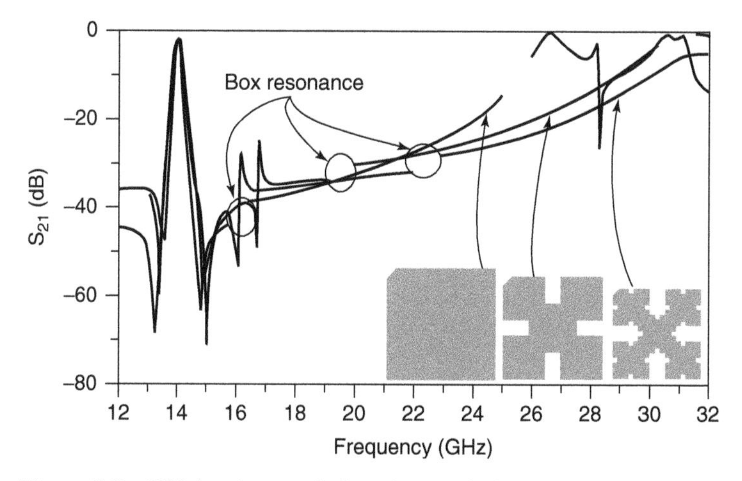

Figure 7.2 Wideband transmission characteristics of dual-mode filters.

resonator. This dissymmetry is very important since it makes it possible to control the coupling between two orthogonal modes. This topology also introduces a transmission zero one each side of the passband, enhancing the selectivity of the filter. The filters in Figure 7.1 are therefore second-order filters with two transmission zeros. When applying Minkowski's iteration to the resonator, the passband decreases due to weaker couplings, and the transmission zeros get closer to the passband. In Figure 7.1, the insertion loss for the square resonator filter is 1.8 dB, 1.68 dB for the first-iteration Minkowski filter, and 1.85 dB for the second-iteration Minkowski filter.

Figure 7.2 shows the wideband transmission characteristics of the three dual-mode filters. It can be seen that the attenuation at twice the center frequency increases when the fractal iteration number increases. Another advantage is that the additional miniaturization allows using enclosures with smaller dimensions, thus providing a situation where the resonant frequency of the enclosure will be farther away from the operational bandwidth of the filter.

The design of a dual-mode Minkowski resonator filter requires defining the appropriate prototype circuit to use. For a single dual-mode resonator, one must use a prototype circuit of order 2 with two transmissions zeros, one on each side of the passband. The bandpass prototype that satisfies these requirements is shown in Figure 7.3. Each $L-C$ resonator represents a resonant mode of the dual-mode resonator. The coupling J_1 is negative, to produce a transmission zero on the left side of the passband, while the coupling J_2 is positive, to create the transmission zero on the right side of the passband.

In the prototype, the transmission zero frequency on the left of the passband can be controlled by varying J_1 without changing the location of the transmission zero on the right side of the passband, as shown in Figure 7.4. The transmission zero on the right of the passband can be controlled by varying J_2 as shown in Figure 7.5. For a dual-mode Minkowski filter, adjusting the response is not that

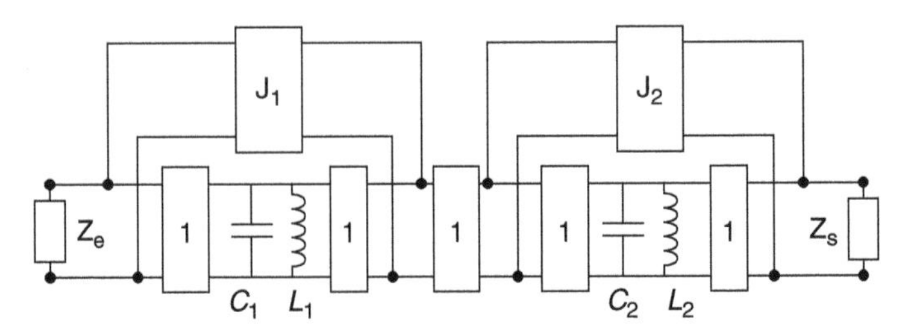

Figure 7.3 Second-order bandpass prototype with two transmission zeros.

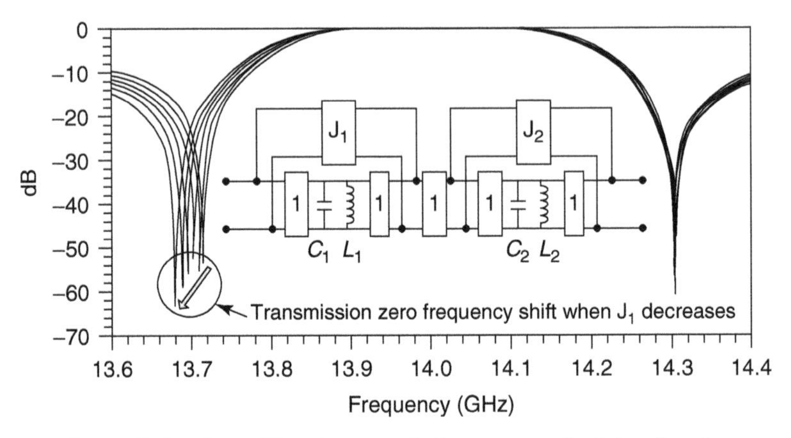

Figure 7.4 Controlling the transmission zero on the left using J_1.

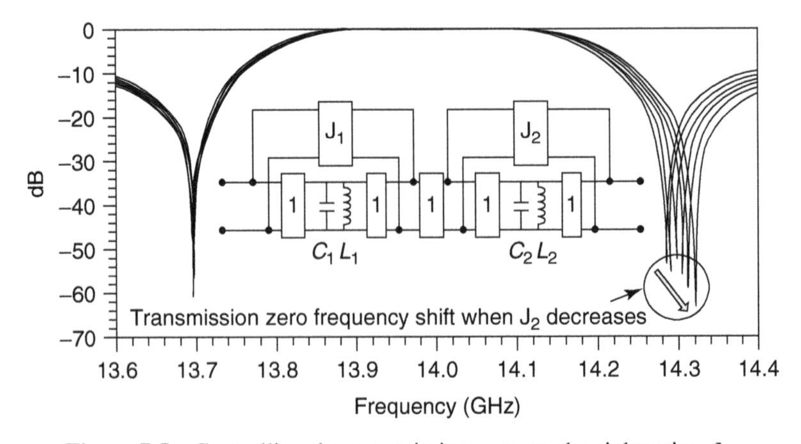

Figure 7.5 Controlling the transmission zero on the right using J_2.

simple. The location of the transmission zeros depend on the couplings between the two resonant modes and the input/output coupling. Referring to Figure 7.6, adjusting J_1 can be done by changing the coupling between the input line and resonant mode 2, and adjusting J_2 can be done by changing the coupling between the output line and resonant mode 1. The input and output couplings d_{in} and d_{out} are used to control the external quality factor.

Figures 7.7 and 7.8 show how the locations of the transmission zeros change when the input and output couplings are changed. These figures show that the

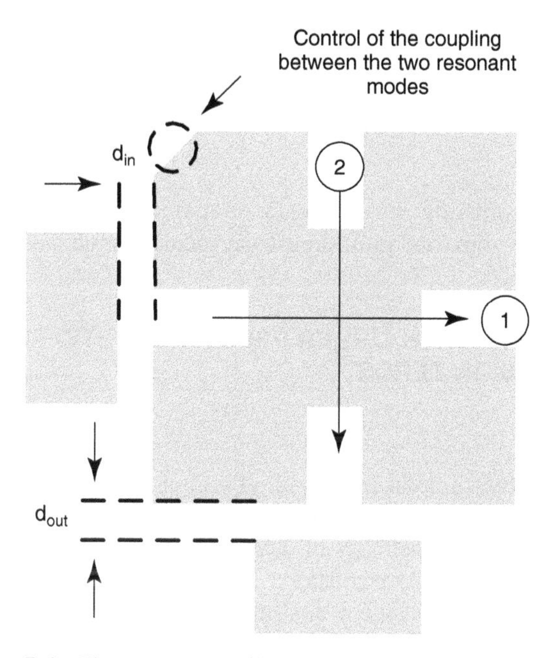

Figure 7.6 Elements responsible for adjusting the filter response.

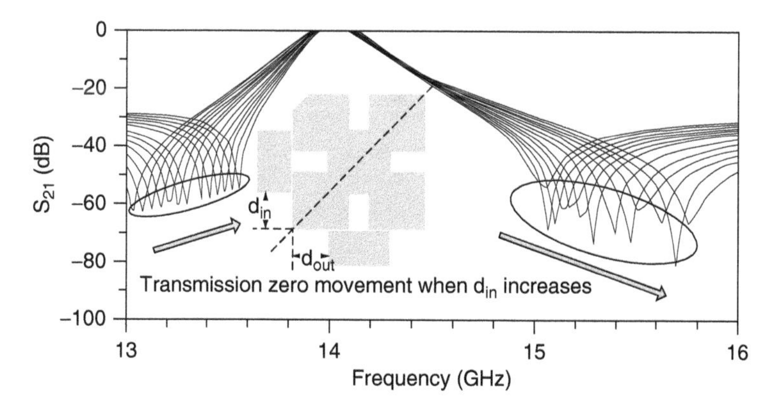

Figure 7.7 Effect of d_{in} on the location of the transmission zeros.

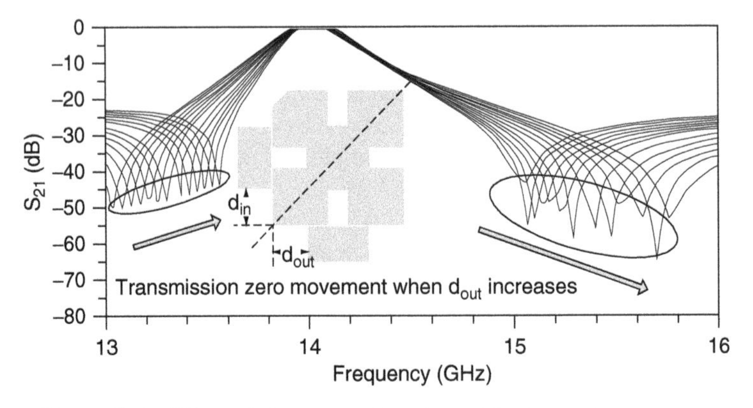

Figure 7.8 Effect of d_{out} on the location of the transmission zeros.

location of the transmission zeros depends on more than one parameter and that when adjusting one zero, one must adjust the second simultaneously.

7.3 DESIGN OF FOURTH-ORDER PSEUDO-ELLIPTIC FILTERS WITH TWO TRANSMISSION ZEROS

7.3.1 Topology

The topology of a fourth-order pseudo-elliptic filter with two transmission zeros using dual-mode Minkowski resonators is presented in Figure 7.9. The topology was inspired by the filter topologies of dual-mode filters of Chapter 1. It consists of two dual-mode Minkowski resonators. The coupling between two orthogonal resonant modes (1 and 2) inside a dual-mode resonator is achieved by cutting off one of the resonator corners. This introduces a dissymmetry in the resonator that allows separating the resonant frequencies of the two orthogonal modes. The additional bended line provides a coupling between modes 2 and 3. The

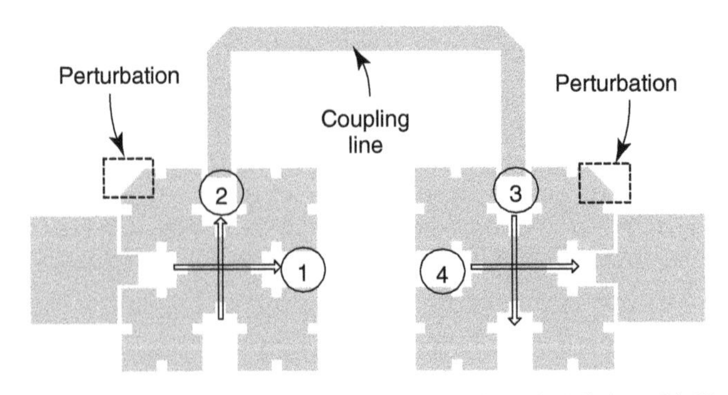

Figure 7.9 Topology of the fourth-order dual-mode Minkowski filter.

coupling between modes 1 and 4 depends on the distance separating the two resonators. This coupling can create two transmission zeros, one on each side of the passband. The external quality factor is determined by the coupling between the input line and resonant mode 1, or the coupling between the resonant mode 4 and the output line. The input and output lines have a 50-Ω impedance to connect to the external system and excite polarizations 1 and 4.

7.3.2 Synthesis

The in-line bandpass prototype of Chapter 2 must be modified slightly to account for the expected behavior of the dual-mode Minkowski filter of Figure 7.9. The modified prototype is shown in Figure 7.10. The four L–C resonators correspond to the fourth order of the filter, and the cross-coupling J between resonators 1 and 4 is used to create a transmission zero on each side of the passband [2–4]. The synthesis method presented in Chapter 2 does not account for the case of the in-line prototype of Figure 7.10. From Chapter 2 there should be a cross-coupling J between resonators 1 and 3 and/or between resonators 2 and 4. Therefore, a new synthesis method is needed and is presented below.

To illustrate the method we use the example of a Ku band filter used in a satellite repeater [2] with a specified center frequency of 14.125 GHz and a bandwidth of 250 MHz. The synthesis method starts by performing the synthesis of a fourth-order Chebyshev bandpass prototype circuit centered at 14.125 GHz with a 250-MHZ bandwidth. Figure 7.11 shows the bandpass prototype and element values.

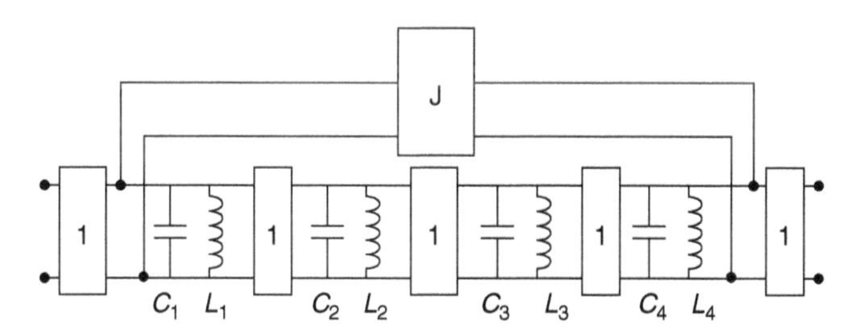

Figure 7.10 Fourth-order pseudo-elliptic bandpass prototype with two symmetrical transmission zeros.

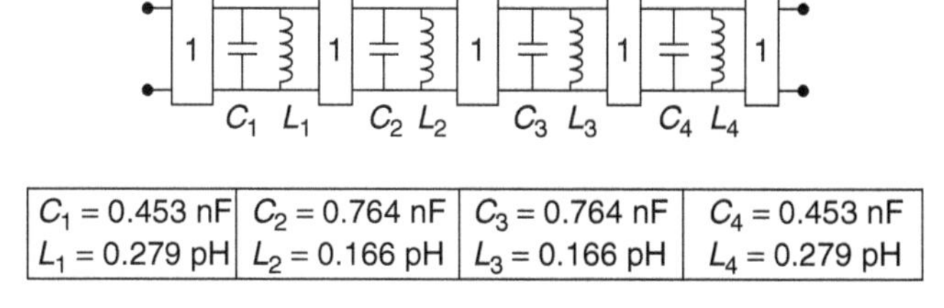

$C_1 = 0.453$ nF	$C_2 = 0.764$ nF	$C_3 = 0.764$ nF	$C_4 = 0.453$ nF
$L_1 = 0.279$ pH	$L_2 = 0.166$ pH	$L_3 = 0.166$ pH	$L_4 = 0.279$ pH

Figure 7.11 Fourth-order Chebyshev bandpass prototype.

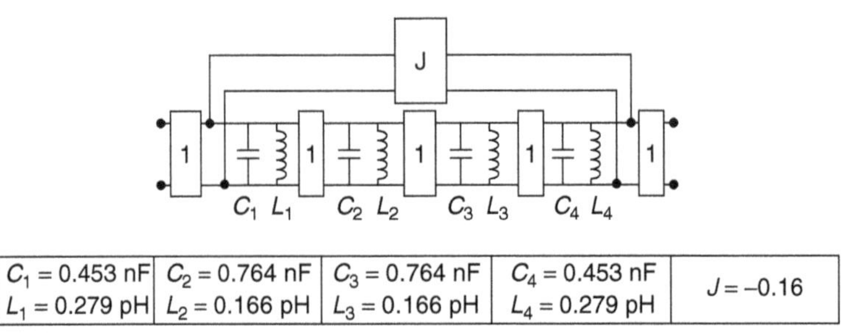

$C_1 = 0.453$ nF	$C_2 = 0.764$ nF	$C_3 = 0.764$ nF	$C_4 = 0.453$ nF	$J = -0.16$
$L_1 = 0.279$ pH	$L_2 = 0.166$ pH	$L_3 = 0.166$ pH	$L_4 = 0.279$ pH	

Figure 7.12 Fourth-order pseudo-elliptic bandpass prototype with two symmetrical transmission zeros.

The cross-coupling J is then added to introduce the transmission zeros. Preserving the initial prototype element values (L_i, C_i), it is possible to place the transmission zeros correctly without degrading the passband characteristics by adjusting the value of J. For example, when $J = -0.16$, the transmission zeros are placed symmetrically around the passband at 13.82 and 14.4 GHz. Figure 7.12 summarizes the values of the pseudo-elliptic bandpass prototype.

Figure 7.13 shows the response of a fourth-order Chebyshev and the fourth-order pseudo-elliptic bandpass prototype obtained previously. It can be seen that the passband characteristics have generally been preserved, while the selectivity of the filter has been improved by the introduction of two transmission zeros, one on each side of the passband. The theoretical coupling coefficients, computed from the prototype element values using the equations of Hong and Lancaster [5], are $K_{12} = 0.0151$, $K_{23} = 0.0139$ and $K_{34} = 0.0151$, $K_{14} = -0.0039$. The next steps consist of defining the physical parameters of the filter to achieve the theoretical coupling coefficients.

7.3.3 Design Steps

Bent Line The bent line that goes from the first to the second dual-mode resonator is used to couple vertical resonant modes 1 and 3, as shown in Figure 7.14. The value of the coupling depends on the line as well as on the distance d_{13} between the center of the resonator and the tip of the bent line. This coupling should be made equal to the prototype theoretical coupling $K_{23} = 0.0139$.

To calculate the length of a line that can create a positive coupling, we consider the equivalence between coupling from mutual inductance and a lossless transmission line of characteristic impedance Z_0 coupling two resonators of respective characteristic impedances Z_1 and Z_2, as shown in Figure 7.15. The *ABCD* matrix of a lossless transmission line of length L and characteristic impedance Z_0 is given by

$$\begin{bmatrix} A & B \\ C & D \end{bmatrix} = \begin{bmatrix} \cos\beta L & jZ_0 \sin\beta L \\ jY_0 \sin\beta L & \cos\beta L \end{bmatrix}$$

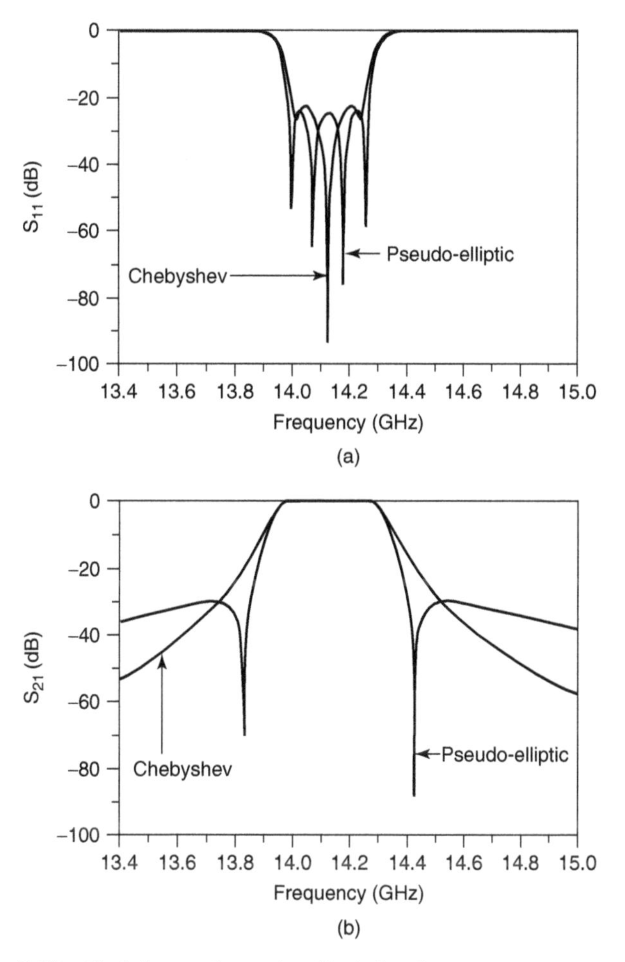

Figure 7.13 Chebshev and pseudo-elliptic bandpass prototype responses.

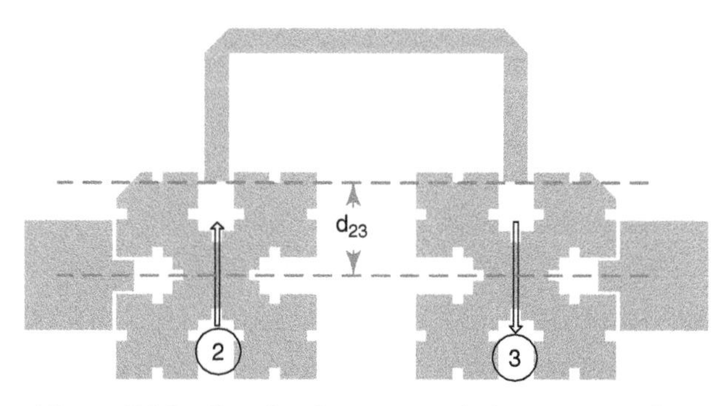

Figure 7.14 Coupling between vertical resonant modes.

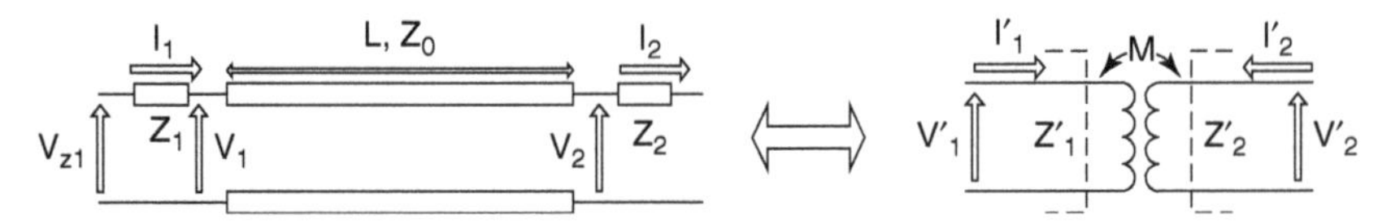

Figure 7.15 Equivalence between the coupling created by a transmission line and the mutual coupling between two resonators.

where β is the propagation constant ($\beta = 2\pi/\lambda_g$, λ_g the guided wavelength) and $Y_0 = 1/Z_0$. Since by definition

$$\begin{bmatrix} V_2 \\ -I_2 \end{bmatrix} = \begin{bmatrix} A & B \\ C & D \end{bmatrix} \begin{bmatrix} V_1 \\ -I_1 \end{bmatrix}$$

we get

$$I_2 = -jY_0 \sin(\beta L) V_1 + \cos(\beta L) I_1$$

$$V_1 = -j\frac{\cot \beta L}{Y_0} I_1 + j\frac{1}{Y_0 \sin \beta L} I_2$$

$$V_{z1} = V_1 + Z_1 I_1 \Leftrightarrow V_{z1} = \left(-j\frac{\cot \beta L}{Y_0} + Z_1\right) I_1 + \left(j\frac{1}{Y_0 \sin \beta L}\right) I_2$$

On the other hand, the impedance matrix equation of the mutual inductance (ideal transformer) between two resonators is given by

$$\begin{bmatrix} V_1' \\ V_2' \end{bmatrix} = \begin{bmatrix} Z_1' & j\omega M \\ j\omega M & Z_2' \end{bmatrix} \begin{bmatrix} I_1' \\ I_2' \end{bmatrix} \Leftrightarrow V_1' = Z_1' I_1' + j\omega M I_2'$$

Since $I_1 = I_1'$ and $I_2 = -I_2'$, when comparing the two input voltages we have

$$Z_1' = Z_1 - j\frac{\cot \beta L}{Y_0} \quad \text{and} \quad M = -\frac{Z_0}{\omega \sin \beta L}$$

In the same manner, we find that

$$Z_2' = Z_2 - j\frac{\cot \beta L}{Y_0}$$

For the impedances to be the same, we need $\cot \beta L = 0$. This leads to

$$L = (2k+1)\frac{\lambda_g}{4} \quad \text{and} \quad M = (-1)^{k+1}\frac{Z_0}{\omega}$$

Since the prototype theoretical coupling is positive, k must be odd. By taking $k = 1$, we finally get

$$L = \frac{3\lambda_g}{4} \quad \text{and} \quad M = \frac{Z_0}{\omega}$$

For the line to create a positive coupling, the smallest value for the line length is $L = 3\lambda_g/4 = 14.49$ mm.

Enclosure Design A first estimation of the gap spacing between resonators is used to define the enclosure length, which needs to be $L = 20$ mm. The width should be large enough to prevent parasitic couplings with the enclosure walls. In addition, when the length is known, the resonant frequencies of various modes can be computed in terms of the length/width ratio $R = L/l$. These are shown in Figure 7.16. We note that for R varying between 1 and 1.32, and for f varying between 12 and16 GHz, there are no propagation modes in this zone. Taking $R = 1.176$, for example, provides an enclosure width of 17 mm. In this case, the enclosure resonances occur at 11.2 and 16.8 GHz and surround fully the operational band of the filter centered at 14.125 GHz.

Matching the Coupling Coefficients and External Quality Factor In the first step the gap distance between the input line and the first resonator, or the gap distance between last resonator and output line, is defined. The input or output gap is adjusted to match the theoretical external quality factor of the bandpass

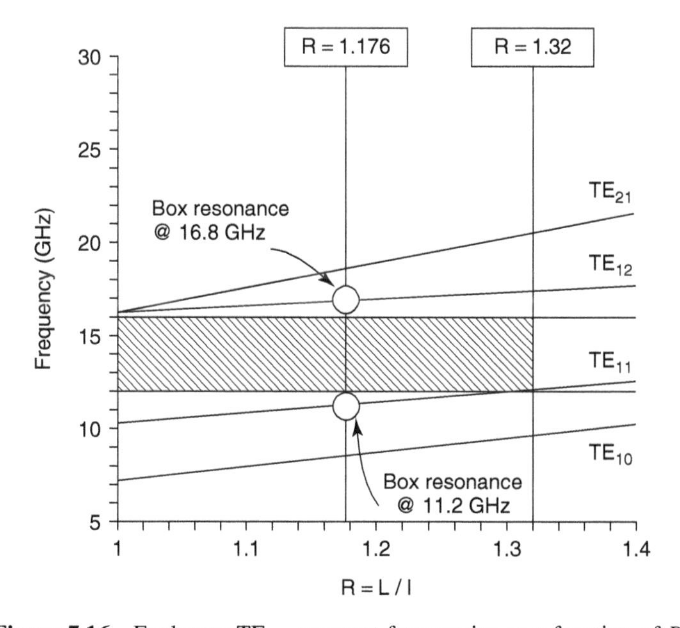

Figure 7.16 Enclosure TE_{nm} resonant frequencies as a function of R.

prototype. The two principal techniques used to define the external quality factor of the dual-mode Minkowski filter are described below.

Knowing that a dual-mode resonator has two resonant modes, two configurations are possible: horizontal and vertical polarizations, as shown in Figure 7.17(a). A first technique for defining the external quality factor, shown in Figure 7.17(b), consists of calculating the ratio of the resonant frequency f_0 over the bandwidth Δf corresponding to a $\pm 90^\circ$ phase shift in the reflection coefficient. This can be done with simulations using Agilent's Momentum software. In that case the experimental external quality factor is given by

$$Q_{\text{ext}} = \frac{f_0}{\Delta f}$$

Another method of defining the external quality factor is to consider the transmission characteristic of a resonator with input and output lines as shown in Figure 7.18(a). As can be seen in Figure 7.18(b), the external quality factor is defined by calculating the ratio of the resonant frequency f_0 over $\Delta' f$ corresponding to the -3 dB bandwidth of the transmission coefficient. The experimental

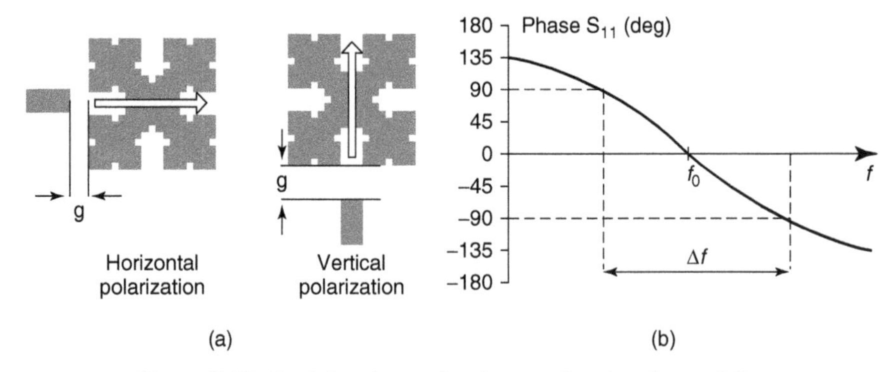

Figure 7.17 Defining the quality factor using the phase of S_{11}.

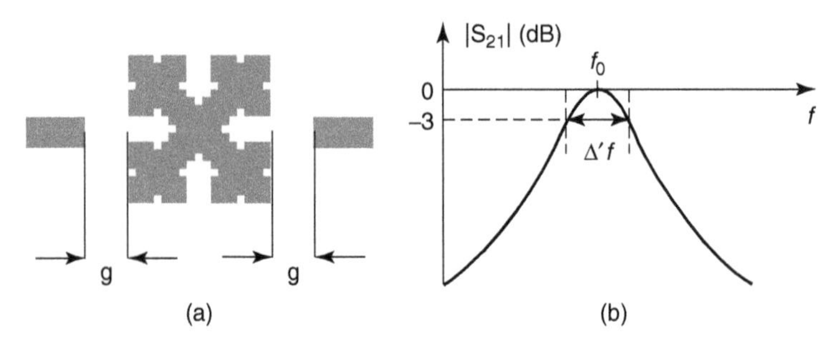

Figure 7.18 Defining the external quality factor using the magnitude of S_{21}.

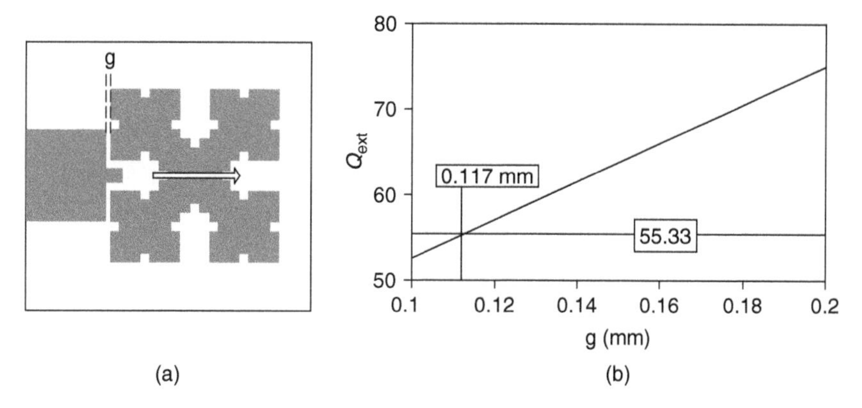

Figure 7.19 Matching the theoretical quality factor.

external quality factor is then given by

$$Q_{\text{ext}} = 2Q'_{\text{ext}} = 2\frac{f_0}{\Delta f'}$$

In this case, the resonator is excited by two 50-Ω characteristic impedance lines. The required line width is calculated using Agilent's LineCalc. The characteristic impedance depends on the cavity heights, thickness, and permittivity of the substrate. For example, using the phase characteristic, Figure 7.19 shows how the external quality factor varies in terms of the gap distance g. For a theoretical external quality factor of $Q_{\text{ext}} = 55.33$ from prototype computations, the gap distance between the input line and the first resonator should be taken as $g = 0.117$ mm.

The second step consists of defining the gap distance d_{14} between the two resonators, corresponding to the coupling coefficient K_{14}. This is done by simulations of the setup in Figure 7.20. The input/output lines are placed sufficiently far away from the resonators so that they do not interfere with the value of the coupling K_{14} between resonators. Figure 7.21(a) shows how the resonant frequencies of the resonators vary when the distance between resonators increases. Figure 7.21(b) shows how the coupling coefficient K_{14} changes when the distance between resonators varies. Using the curve, one needs a distance $d_{14} = 2.684$ mm to achieve the desired theoretical coupling $K_{14} = 0.0039$.

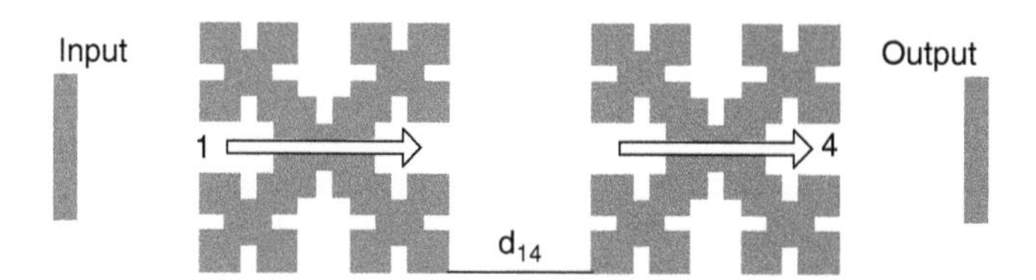

Figure 7.20 Setup for calculating the coupling k_{14}.

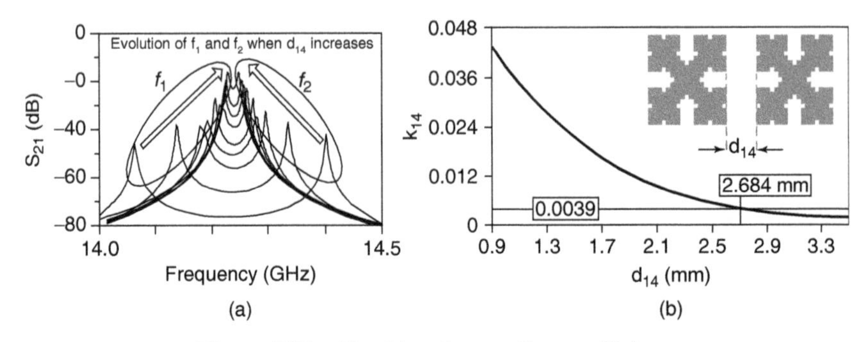

Figure 7.21 Matching the coupling coefficient.

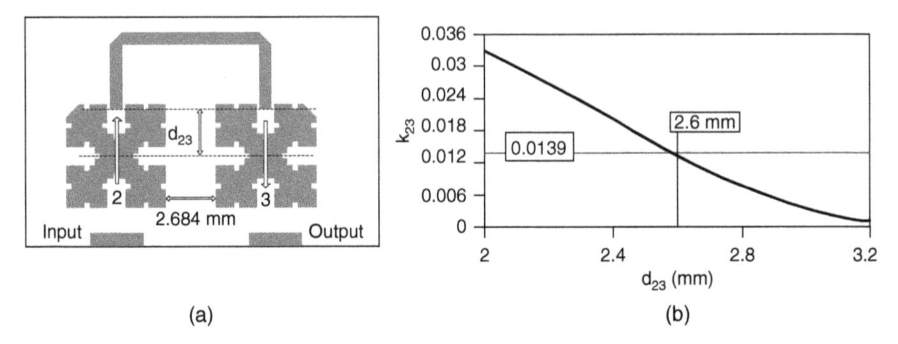

Figure 7.22 Matching the coupling coefficient k_{23}.

The next step consists of defining the gap distance d_{23} to achieve the desired coupling coefficient K_{23}. The gap d_{23} corresponds to the distance between the middle of the resonator and the edge of the bent line as shown in Figure 7.22(a). Figure 7.22(b) shows how coupling between modes 2 and 3 varies when d_{23} is changed. From the curve, the gap distance should be $d_{23} = 2.6$ mm to achieve the theoretical coupling coefficient $K_{23} = 0.0139$.

Finally, the coupling K_{12} between the two resonant modes 1 and 2 inside the first dual-mode resonator is achieved by adjusting the cut a on the corner of the resonator as shown in Figure 7.23(a). The same technique is used to achieve the coupling K_{34} between the two resonant modes 3 and 4 inside the second dual-mode resonator. Figure 7.23(b) shows how the coupling between modes 1 and 2 varies by changing the depth a of the cut. From the curve, the depth of the cut should be taken as $a = 0.48$ mm to achieve the theoretical coupling $K_{12} = 0.0139$.

Final Adjustments It is clear that the design method explained earlier relied on simulations of the different elements of the filter taken in isolation from the rest of the structure. Figure 7.24(a), for example, shows the response of the filter using the initial gap distances. One can see that the response is not satisfactory.

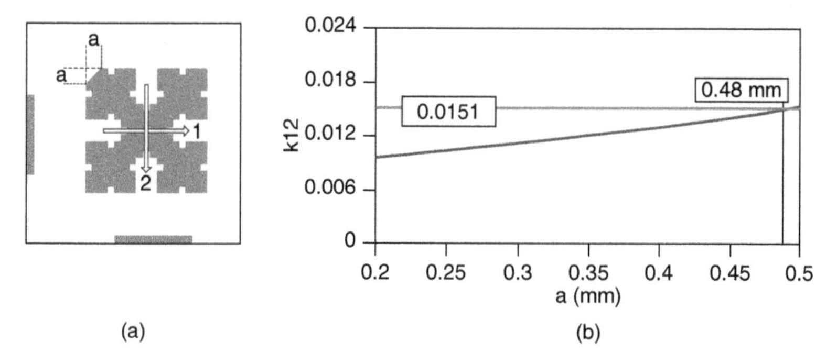

Figure 7.23 Matching the coupling coefficient k_{12}.

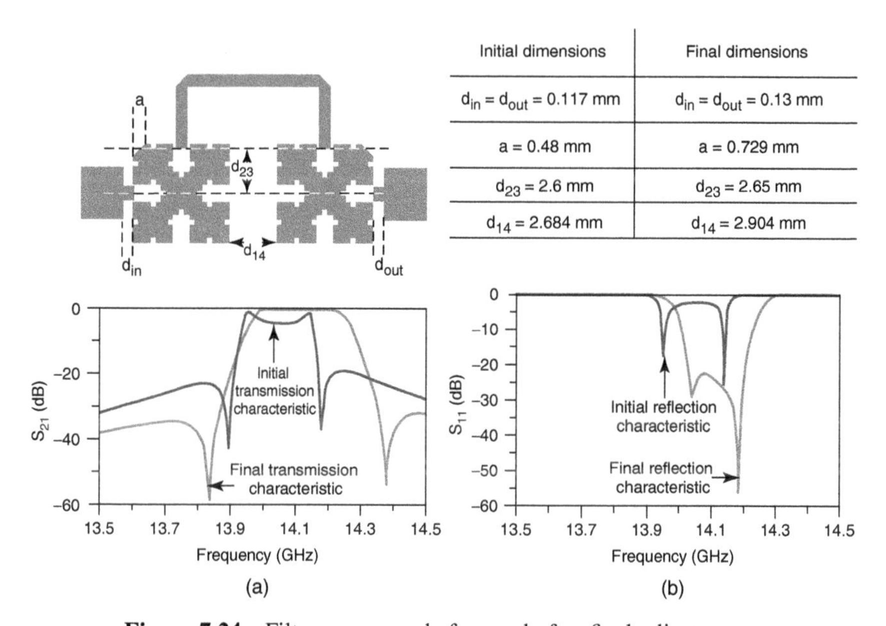

Initial dimensions	Final dimensions
$d_{in} = d_{out} = 0.117$ mm	$d_{in} = d_{out} = 0.13$ mm
a = 0.48 mm	a = 0.729 mm
$d_{23} = 2.6$ mm	$d_{23} = 2.65$ mm
$d_{14} = 2.684$ mm	$d_{14} = 2.904$ mm

Figure 7.24 Filter responses before and after final adjustments.

Therefore, additional adjustments are needed using simulations of the entire filter so that the various intercoupling effects are also accounted for. After some trial and error, the final gap values and corresponding final filter responses are given in Figure 7.24(b).

Figure 7.25 shows the wideband transmission characteristic of the fourth-order pseudo-elliptic filter with two transmission zeros using dual-mode Minkowski resonators. The enclosure resonances calculated by Momentum occur at 11.2 and 16.8 GHz. A third transmission zero appears at 15.64 GHz and the bent line coupling resonates at 16.35 GHz. The attenuation at the transmission zero on the left of the passband is superior to 34.8 dB, and the attenuation at the transmission zero on the right of the passband is greater than 32.2 dB.

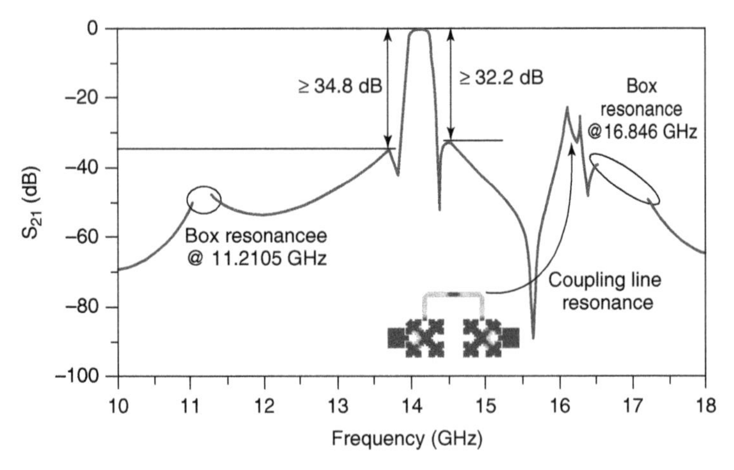

Figure 7.25 Simulated wideband characteristics of the dual-mode Minkowski filter.

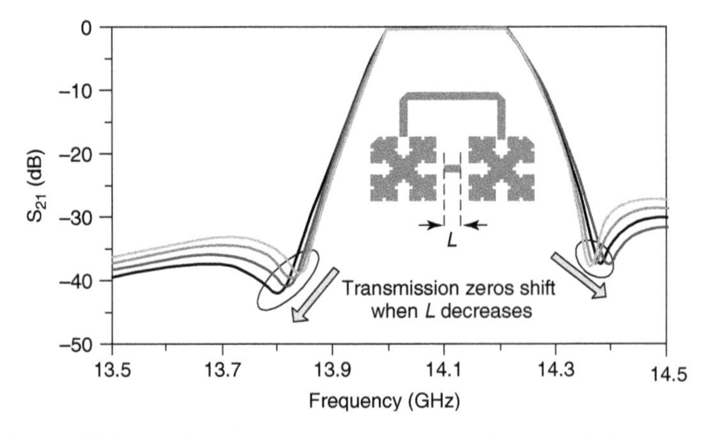

Figure 7.26 Additional coupling line to adjust the transmission zeros.

As mentioned earlier, the coupling coefficient K_{14} determined by the gap d_{14} controls the positions of the transmission zeros on both sides of the passband. It is possible to control K_{14} further by adding a small coupling line of length L between the two dual-mode resonators. Figure 7.26 shows how this additional coupling line changes the positions of the transmission zeros without degrading the rest of the response. When L decreases, the coupling coefficient K_{14} decreases and the transmission zeros are shifted away from the passband.

7.4 REALIZATION AND MEASURED PERFORMANCE

The fourth-order pseudo-elliptic Minkowski filter presented in previous sections was constructed and its performance was measured. The enclosure had a width of

17 mm and a length of 20 mm. Duroid substrate with a relative dielectric permittivity of 2.2 and a thickness of 0.254 mm was used. The upper and lower cavity heights were 1 mm. The gaps between resonators are shown in Figure 7.27. Fabrication followed the same procedures as in Section 5.2.6. The enclosure was built in-house and the substrate and planar resonators (the filter) were fabricated by Atlantec. The fourth-order pseudo-elliptic filter with two transmission zeros using dual-mode Minkowski resonators that was constructed is shown in Figure 7.28.

The measured and simulated wideband transmission characteristics are included in Figure 7.29, and the measured and simulated reflection characteristics in the vicinity of the passband are given in Figure 7.30. The simulations in this case take dielectric and metallic losses into account. Table 7.1 summarizes the key features of the measured and simulated filters.

We note that the measured characteristics are shifted toward higher frequencies. As for the realizations in Chapters 5 and 6, this can be explained by cavity

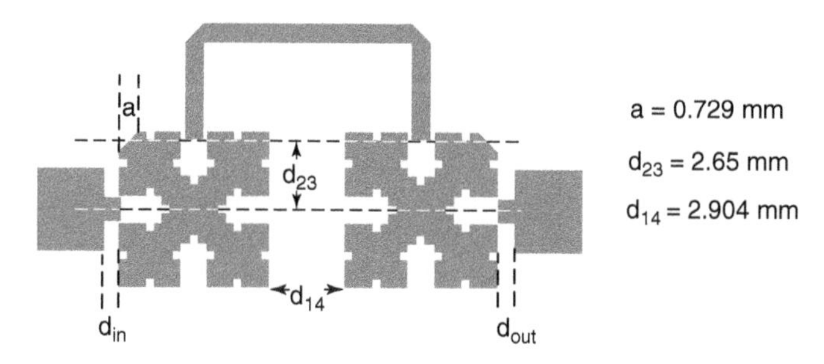

Figure 7.27 Fourth-order pseudo-elliptic dual-mode Minkowski filter.

Figure 7.28 Realization of the fourth-order pseudo-elliptic dual-mode Minkowski filter.

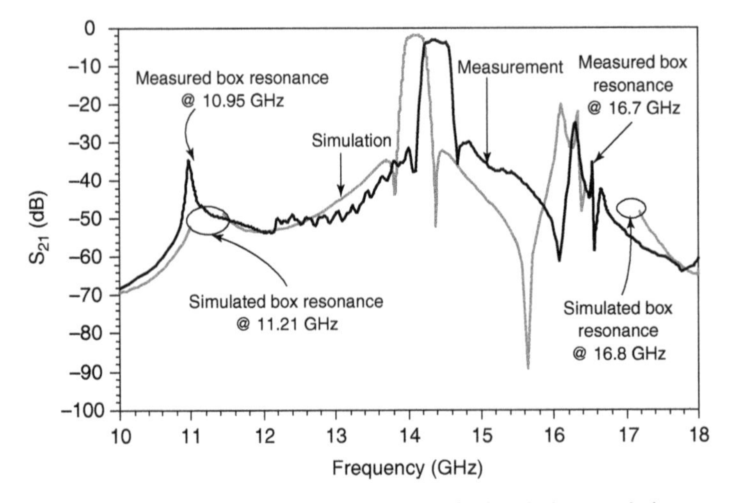

Figure 7.29 Measured and simulated wideband characteristics.

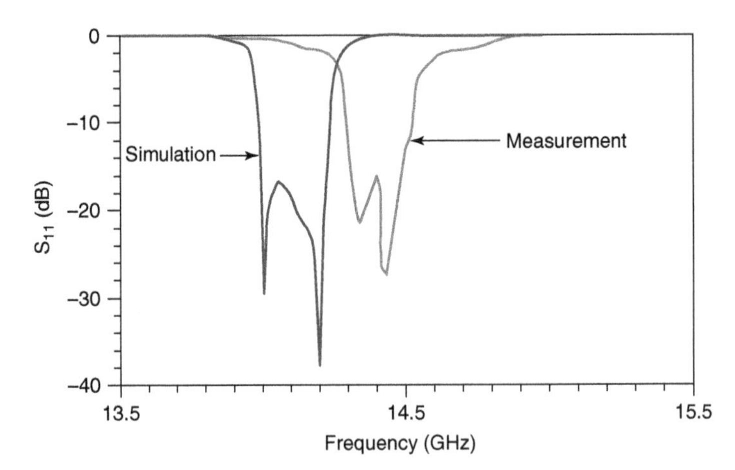

Figure 7.30 Measured and simulated return characteristics in the vicinity of the passband.

TABLE 7.1 Summary of Key Filter Characteristics

Filter Characteristic	Simulation	Measurement
Central frequency	14.125 GHz	14.38 GHz
Bandwidth	250 MHz	352 MHz
Insertion loss	−2 dB	−3.1 dB
Return loss	≤−17 dB	≤−15 dB
Zero 1	13.82 GHz	14.09 GHz
Zero 2	14.38 GHz	14.68 GHz
First box resonance	11.21 GHz	10.95 GHz
Second box resonance	16.8 GHz	16.7 GHz

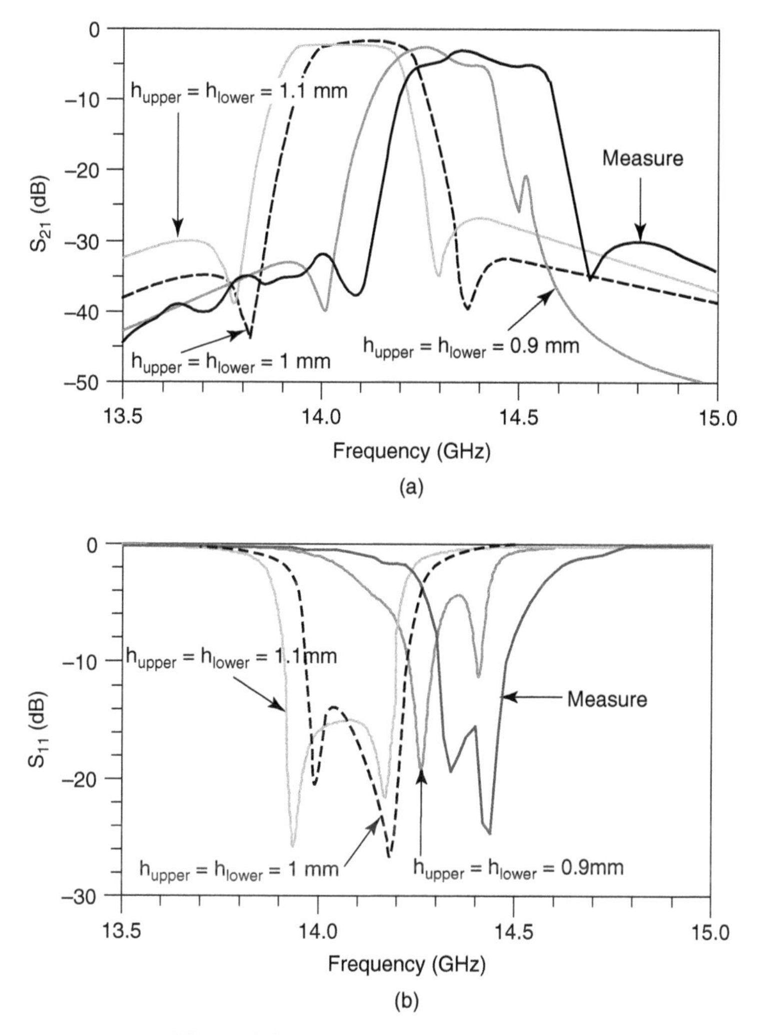

Figure 7.31 Influence of cavity heights.

heights greater than expected, substrate thinner than expected, and the use of low mesh density during simulations. Figure 7.31 shows how the response of the filter changes when the cavity heights are modified slightly to represent fabrication errors. Compared to the case of the filters in Chapters 5 and 6, the transmission characteristics shift toward higher frequencies when the cavity height decreases.

Figure 7.32 shows how the filter response changes when the substrate thickness is modified. The propagation characteristics shift to higher frequencies when the substrate thickness decreases. Figure 7.33 shows the propagation characteristics for various mesh densities. The propagation characteristics shift toward higher

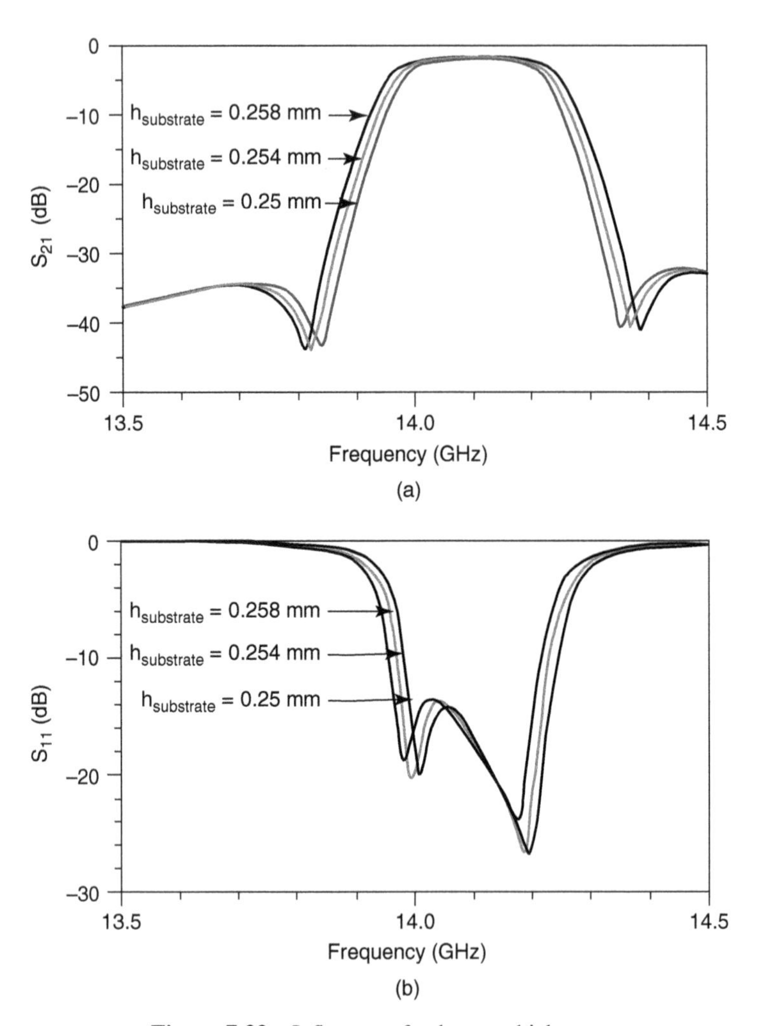

Figure 7.32 Influence of substrate thickness.

frequencies when the mesh density increases. On the other hand, the simulation time is increased when the mesh density increases.

SMA connectors have nonnegligible losses for very high frequencies. In Figure 7.34 a lossy SMA connector is modeled as a 50-Ω impedance in parallel with a parasitic capacitance. Figure 7.35 provides simulations of the filter response when a model of a lossy SMA connectors is used. It can be seen that the transmission characteristics shift toward lower frequencies when introducing the parasitic capacitances. Severe degradations in the return loss of the filter are observed.

The measurement results show good agreement with simulations except for a frequency shift in the transmission characteristics. Using retro-simulations, we can suppose that the actual substrate was thinner than specified and the cavity

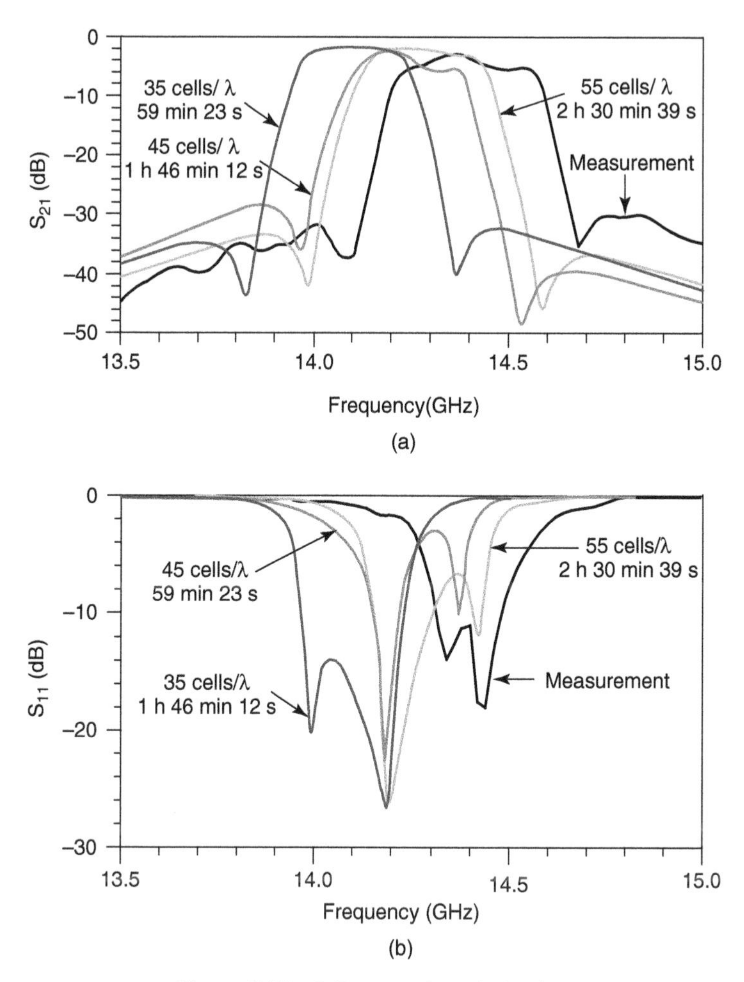

Figure 7.33 Influence of mesh density.

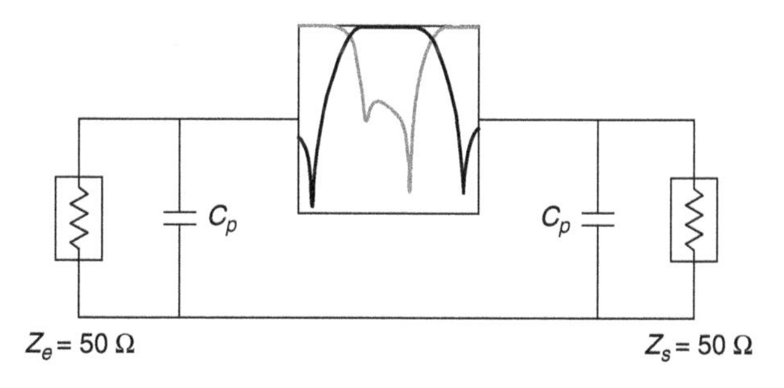

Figure 7.34 Model used for lossy SMA connectors.

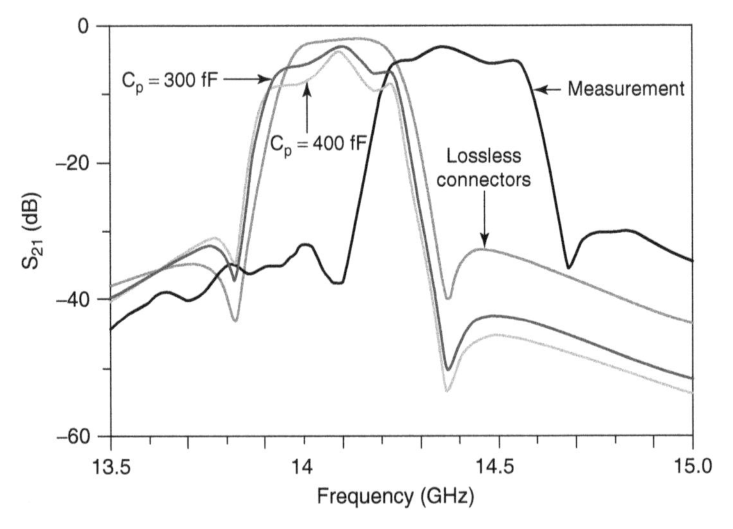

Figure 7.35 Influence of SMA connectors.

heights were larger than specified. For precise simulation results, a high mesh density is required even if the simulation time increases. Moreover, SMA connectors can introduce additional losses not accounted for in the simulations. For better results at high frequencies, one may use K connectors instead of the usual SMA connectors.

REFERENCES

[1] M. Guglielmi, Microstrip ring resonator dual-mode filters, *Workshop on Microwave Filters for Space Applications*, ESA/ESTEC, June 1990.

[2] E. Hanna, P. Jarry, E. Kerhervé, and J. M. Pham, A design approach for capacitive gap parallel-coupled line filters and fractal filters with the suspended substrate technology, in *Microwave Filters and Amplifiers* (scientific editor P. Jarry), Research Signpost, Tivandrum, India, pp. 49–71, 2005.

[3] E. Hanna, P. Jarry, E. Kerhervé, and J. M. Pham, A novel compact dual-mode bandpass filter using fractal shaped resonators, *13th IEEE International Conference on Electronics Circuits and Systems* (*ICECS 2006*), University of Nice, Nice, France, pp. 343–346, Dec. 10–13, 2006.

[4] E. Hanna, P. Jarry, E. Kerhervé, and J. M. Pham, Miniaturization of dual-mode resonators by fractal iteration for suppression of second harmonic, Special Issue on RF and Microwave Filters, Modelling and Design (invited editors P. Jarry and H. Baher), *International Journal for RF and Microwave Computed-Aided Engineering*, Vol. 17, No. 1, pp. 41–48, Jan. 2007.

[5] J. S. Hong and M. J. Lancaster, *Microstrip Filters for RF/Microwave Applications*, Wiley, Hoboken, NJ, 2001.

APPENDIX 1

EQUIVALENCE BETWEEN *J* AND *K* LOWPASS PROTOTYPES

An admittance inverter can be modeled as a π circuit, as shown in Figure A1.1. The chain matrix of a π circuit made of admittances is given in Figure A1.2. The chain matrix of the ideal admittance inverter is given in Figure A1.3.

We replace the admittance inverter by its equivalent circuit in the fundamental element of the in-line prototype to enable us to define its transfer function, as shown in Figure A1.4 [1]. The even-mode admittance is obtained by placing an open circuit in the symmetry plane as shown in Figure A1.5. The even-mode admittance is therefore

$$Y_e = jJ + \frac{1}{(C/2)s + j(B/2)}$$

The odd-mode admittance is obtained by placing a short circuit in the symmetry plane, as shown in Figure A1.6. The odd-mode admittance is then

$$Y_o = -jJ$$

Design and Realizations of Miniaturized Fractal RF and Microwave Filters,
By Pierre Jarry and Jacques Beneat

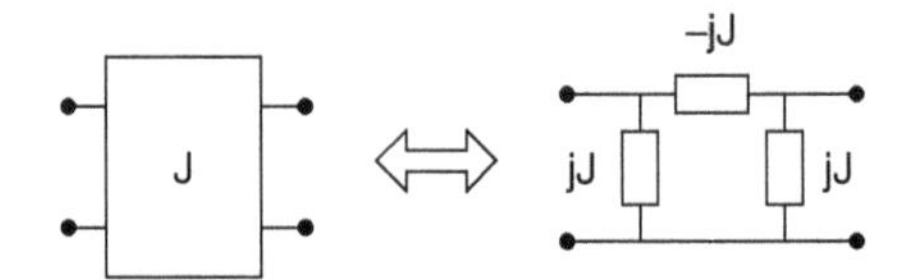

Figure A1.1 Equivalent circuit of an admittance inverter.

$$\begin{pmatrix} A & B \\ C & D \end{pmatrix} = \begin{pmatrix} 1+\dfrac{Y_2}{Y_3} & \dfrac{1}{Y_3} \\ Y_1+Y_2+\dfrac{Y_1 Y_2}{Y_3} & 1+\dfrac{Y_1}{Y_3} \end{pmatrix}$$

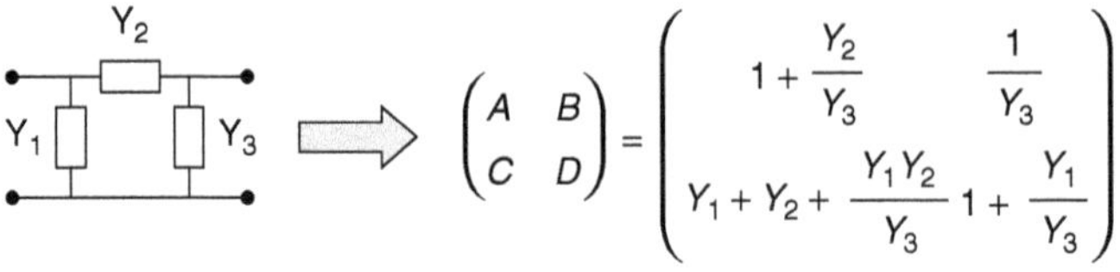

Figure A1.2 Chain matrix of a π circuit.

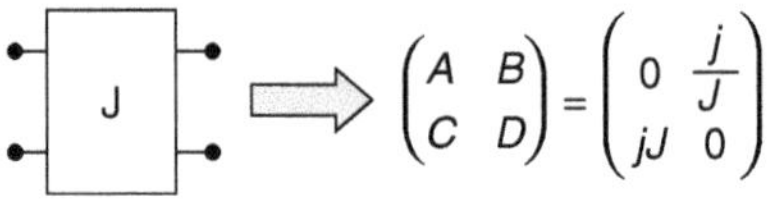

$$\begin{pmatrix} A & B \\ C & D \end{pmatrix} = \begin{pmatrix} 0 & \dfrac{j}{J} \\ jJ & 0 \end{pmatrix}$$

Figure A1.3 Chain matrix of an admittance inverter.

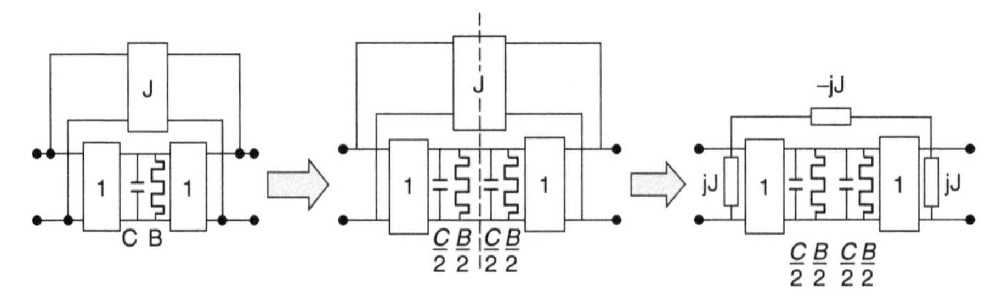

Figure A1.4 Equivalent circuit of the fundamental element.

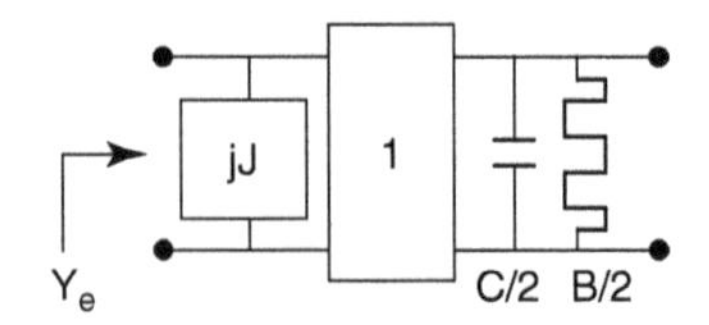

Figure A1.5 Even-mode admittance.

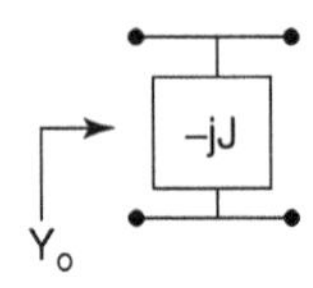

Figure A1.6 Odd-mode admittance.

The transfer function of the fundamental element using even- and odd-mode admittances is given by

$$S_{21}(s) = \frac{Y_e - Y_o}{(1 + Y_e)(1 + Y_o)}$$

$$S_{21}(p) = \frac{jJ(sC + jB) + 1}{(1 - jJ)\left\{1 + (1 + jJ)\left[(C/2)s + j(B/2)\right]\right\}}$$

An impedance inverter can be modeled by the π circuit of Figure A1.7.

The chain matrix of the ideal impedance inverter is shown in Figure A1.8. We replace the impedance inverter by its equivalent circuit in the fundamental element of the in-line prototype to enable us to define its transfer function, as shown in Figure A1.9. The even mode is obtained by placing an open circuit in the symmetry plane, as shown in Figure A1.10. The even-mode admittance is

$$Y_e = \frac{1}{j/K}$$

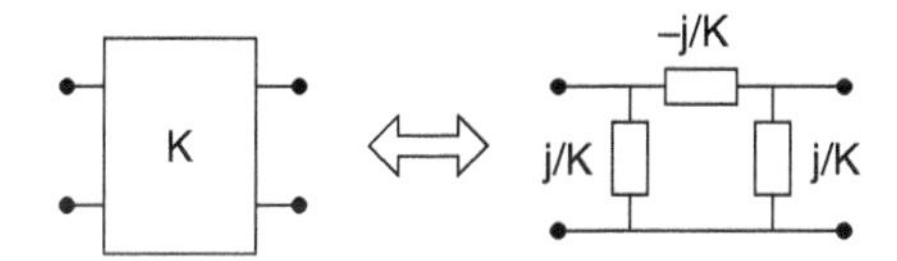

Figure A1.7 Equivalent circuit of an impedance inverter.

$$\begin{pmatrix} A & B \\ C & D \end{pmatrix} = \begin{pmatrix} 0 & jK \\ \frac{j}{K} & 0 \end{pmatrix}$$

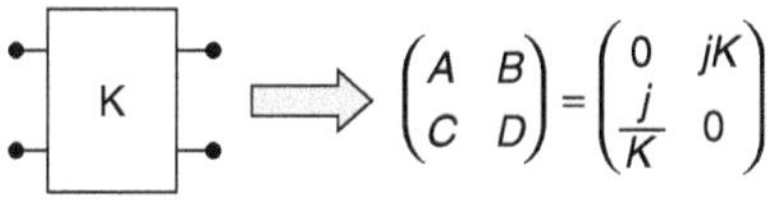

Figure A1.8 Chain matrix of an impedance inverter.

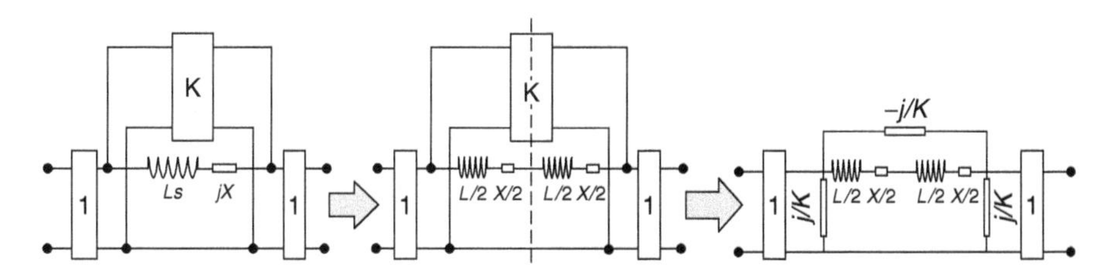

Figure A1.9 Equivalent circuit of the fundamental element.

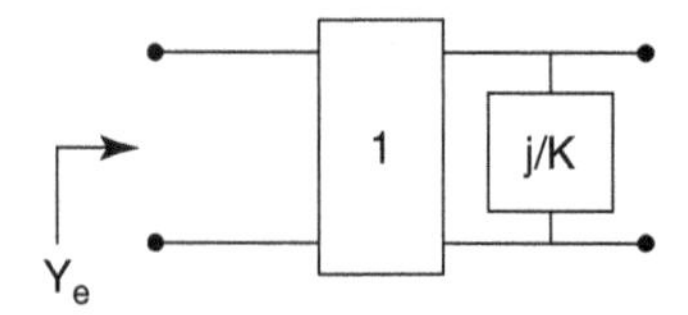

Figure A1.10 Even mode admittance.

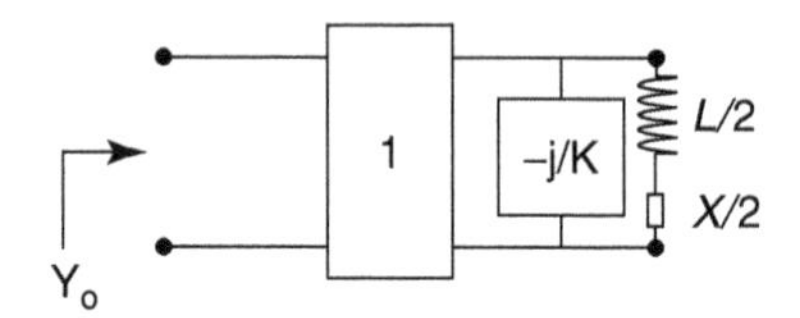

Figure A1.11 Odd-mode admittance.

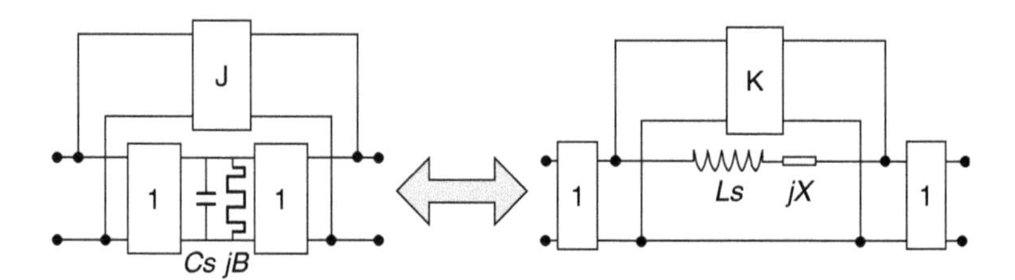

Figure A1.12 Admittance- and impedance-based fundamental elements.

The odd mode is obtained by placing a short circuit in the symmetry plane, as shown in Figure A1.11. The odd-mode admittance is

$$Y_o = \frac{1}{\dfrac{-j}{K} + \dfrac{1}{sL/2 + j(X/2)}}$$

The transfer function of the fundamental element using the even- and odd-mode admittances is given by

$$S_{21}(p) = \frac{Y_e - Y_o}{(1 + Y_e)(1 + Y_o)}$$

$$S_{21}(s) = \frac{1 - (j/K)(Ls + jX)}{\left(1 + \dfrac{j}{K}\right)\left(1 + \left(1 - \dfrac{j}{K}\right)\left(\dfrac{Ls}{2} + \dfrac{jX}{2}\right)\right)}$$

The transfer functions for the fundamental elements based on admittance inverters and on impedance inverters shown in Figure A1.12 will be the same when taking

$$L = C$$

$$X = B$$

$$J = -\frac{1}{K}$$

REFERENCE

[1] P. Jarry, M. Guglielmi, J. M. Pham, E. Kerhervé, O. Roquebrun, and D. Schmitt, Synthesis of dual-modes in-line asymmetric microwave rectangular filters with higher modes, *International Journal of RF and Microwave Computer-Aided Engineering*, pp. 241–248, Feb. 2005.

APPENDIX 2

EXTRACTION OF THE UNLOADED QUALITY FACTOR OF SUSPENDED SUBSTRATE RESONATORS

To extract the unloaded quality factor of suspended substrate resonators, we apply the method developed by Brown *et al*. [1]. Practically, it is impossible to extract the unloaded quality factor of a resonator directly. Only the loaded quality factor can be extracted; the unloaded quality factor is extrapolated. Figure A2.1 shows the setup used to extract the unloaded quality factor of a square dual-mode resonator. The resonator is weakly coupled, so the influence of input/output couplings is minimized.

An attenuation from the maximum of the transmission characteristic of at least 20 dB is needed to neglect the effects of the input/output couplings. An example of the simulation of the transmission characteristic is shown in Figure A2.2. The loaded quality factor of the resonator Q_L is determined using the simulations of the transmission characteristic:

$$Q_L = \frac{f_0}{\Delta f}$$

where f_0 is the resonant frequency and Δf is the -3 dB bandwidth. The unloaded quality factor Q_u is extrapolated from Q_L using the equation

$$Q_u = \frac{Q_L}{1 - S_{21}(f_0)}$$

Design and Realizations of Miniaturized Fractal RF and Microwave Filters,
By Pierre Jarry and Jacques Beneat

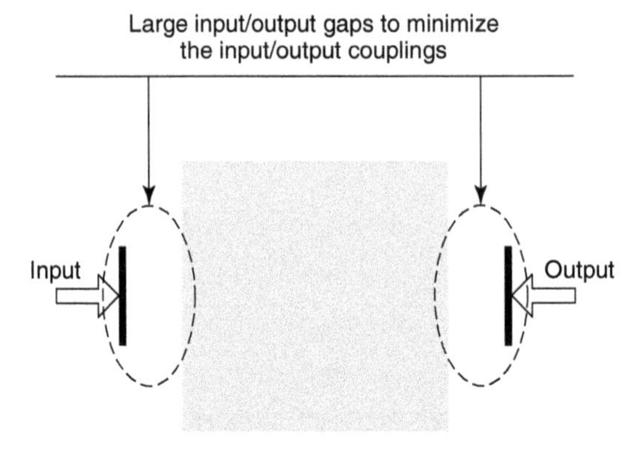

Figure A2.1 Setup for extracting the unloaded quality factor.

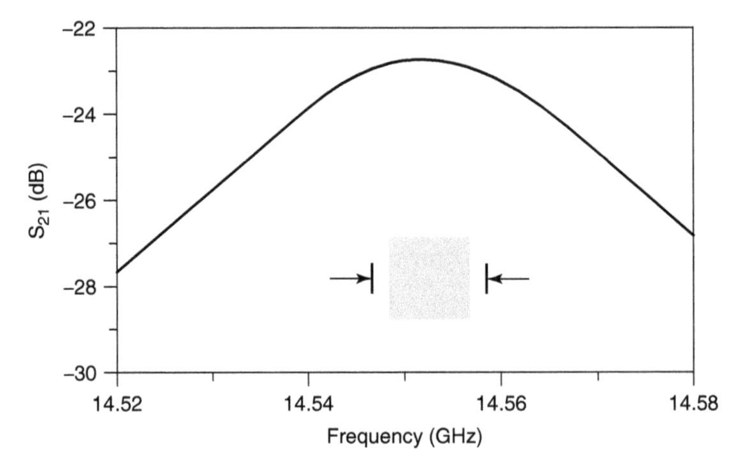

Figure A2.2 Transmission characteristic S_{21} (in dB).

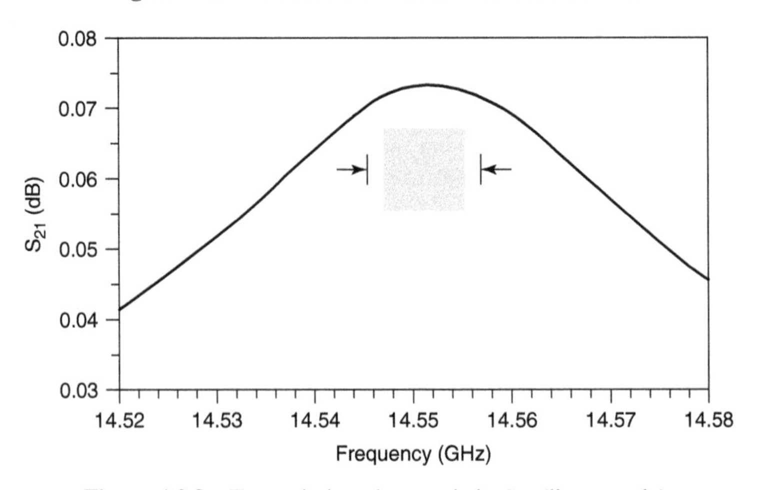

Figure A2.3 Transmission characteristic S_{21} (linear scale).

In this expression, one should use the linear scale value for $S_{21}(f_0)$, as shown in Figure A2.3.

REFERENCE

[1] A. R. Brown, P. Blondy, and G. M. Rebeiz, Microwave and millimeter-wave high-Q micromachined resonators, *International Journal for RF and Microwave Computed-Aided Engineering*, Vol. 9, pp. 326–337, July 1999.

INDEX

ABCD matrix, parameters, 41
Active filters, 1, 22, 23
Algorithm, 49, 51
Amplitude, 4
Approximation, 35
Asymmetric response, 41

Bandpass prototype filter, 102, 103, 118, 167
Bandpass transformation(*s*), 50
Bent line, 5, 168
Bimode, 5

Cantor, 80
Cavity filters, cavity modes, 3
Chebyshev filters, 3, 130, 132, 134, 136
Coplanar filters, 20
Coupled cavity filters, 5
Coupling coefficients, 59, 63, 69, 105, 106, 143, 169, 174, 175
Cross coupled, 41
Curved filters, 5

Dielectric, 4, 9, 123
Dual-mode filters, 85, 161, 162, 163, 166, 176, 177

Electric coupling, 71, 72, 141, 144
E-plane, 12
Evanescent, 3
Even and odd modes, 43, 186
External Q factor, 59, 146

Filter, filtering function, 1, 2
Fractal, 35, 79, 80, 82
Frequency transformation, 50

Gap, 70, 105, 118, 119, 120, 139, 143, 147

Half-wavelength, 81, 93, 102
Hilbert, 92, 93, 129, 130, 133, 134, 136, 145, 154, 155
HTS filters, 23, 24

In-line, 35, 41, 49
Iteration, first and second, 79, 82, 84, 85, 88

Design and Realizations of Miniaturized Fractal RF and Microwave Filters,
By Pierre Jarry and Jacques Beneat

J-inverters, 42, 183, 184

K-inverters, 42, 183, 185

Lowpass filters, prototypes, 36, 103, 183

Magnetic coupling, 71, 72, 108, 140, 148
Meandered line filters, 101, 102, 108, 109, 113, 123
Measured performances, 110, 125, 156, 157, 176
Micro-machined filters, 27
Microstrip filters, 19
Minkowski, 84, 86, 88, 101, 162, 166, 176, 177
Miniaturization, 79, 81
Mono-mode, 74
Multilayer filters, 21, 22

Peano, 80
Planar transmission line filters, 14, 19, 79, 81
Prototype function, 45, 52, 169
Pseudo-elliptic, 37, 49, 101, 136, 137, 139, 144, 145, 147, 149, 154, 166, 167

Quality factor, 3, 59, 120, 131, 138, 142, 172, 173

Return loss, 52, 54, 55, 58, 113, 176

SAW filters, 25, 26
Sensitivity, 83, 90, 96, 97, 98, 114, 125
Siepinski, 80
Superconductivity filters, 23, 24
Suspended substrate structure, 20, 21, 63, 64, 110, 189
Synthesis, 35

Topology, 102, 117, 118, 130, 166
Transfer function, 40

Unloaded quality factor, 66, 89, 91, 94, 95, 96, 189, 190

Von Koch, 80

Zeros of transmission, 44, 73, 102, 103, 107, 113, 117, 122, 123, 137, 140, 144, 145, 147, 149, 154, 164, 165, 166, 167

Lightning Source UK Ltd.
Milton Keynes UK
UKHW021815090819
347717UK00003B/173/P